ELECTRONIC GAMES & TOYS YOU CAN BUILD

by

LEN BUCKWALTER

HOWARD W. SAMS & CO., INC.
THE BOBBS-MERRILL CO., INC.
INDIANAPOLIS • KANSAS CITY • NEW YORK

FIRST EDITION

TWELFTH PRINTING—1975

International Standard Book Number: 0-672-20183-6
Library of Congress Catalog Card Number: 61-17917

PREFACE

Visit a large electronic exhibition which features a giant computer, and you'll find people standing in line to play tick-tack-toe against the electronic "brain." What special fascination attracts both professional engineer and layman to electronic games? The answer, we suspect, is the same one that has made games a popular pastime for 2,000 years—a challenging combination of skill and chance.

What electronics can add to a game or toy is a special "flavor" of its own. Electron currents are invisible, and travel near the speed of light—they can start, stop, split, and control within a brief instant. With these qualities, a game such as the "Electronic Card Trick" managed to confound a group of five engineers who work with these principles daily. The "Mind Reading Act," which uses an extremely simple setup, has consistently created a startling illusion of mental telepathy when performed in front of an audience.

The list lengthens to include projects which will challenge your mental and physical skills. There are gadgets for group entertainment, or for the amusement of children—all easily constructed in an evening or two. In each case, the novelty is made possible by the remarkable performance of an electronic circuit.

In general, the more easily constructed projects are located in the earlier chapters. If you have never constructed an electronic device, you'll do better if you start with one of these. Then you can progress to more difficult ones. Any of the units will provide you with many hours of enjoyment and relaxation, and for a very small cost.

LEN BUCKWALTER

September, 1961

TABLE OF CONTENTS

1. CONSTRUCTION TECHNIQUES

You can build a lot of interesting electronic games and toys with just a piece of board, some simple tools, and a couple of radio parts. Even if you've never built a kit or wired a crystal set, you will find the projects in this book simple to build and fun to use. They progress from easily assembled units in early chapters to more complex ones near the end. If you are a complete novice, try one of the more basic units first. Then, as you gain proficiency in wiring and following instructions, tackle a more advanced circuit. Numerous illustrations—designed to answer such questions as where to place parts, how to wire them together, and how to check out the connections—are sprinkled throughout each chapter.

Schematic diagrams, and explanations of how the circuits operate are included for the benefit of those who are more experienced. Each chapter contains all the information you need to assemble the unit described. Thus, the experienced builder can start anywhere, governed only by his interest and ability.

It's best to read over the complete chapter first. Often a series of steps will be made clear by a statement near the end of a description. The photos and drawings can save you much time by serving as a guide to parts placement and hookup.

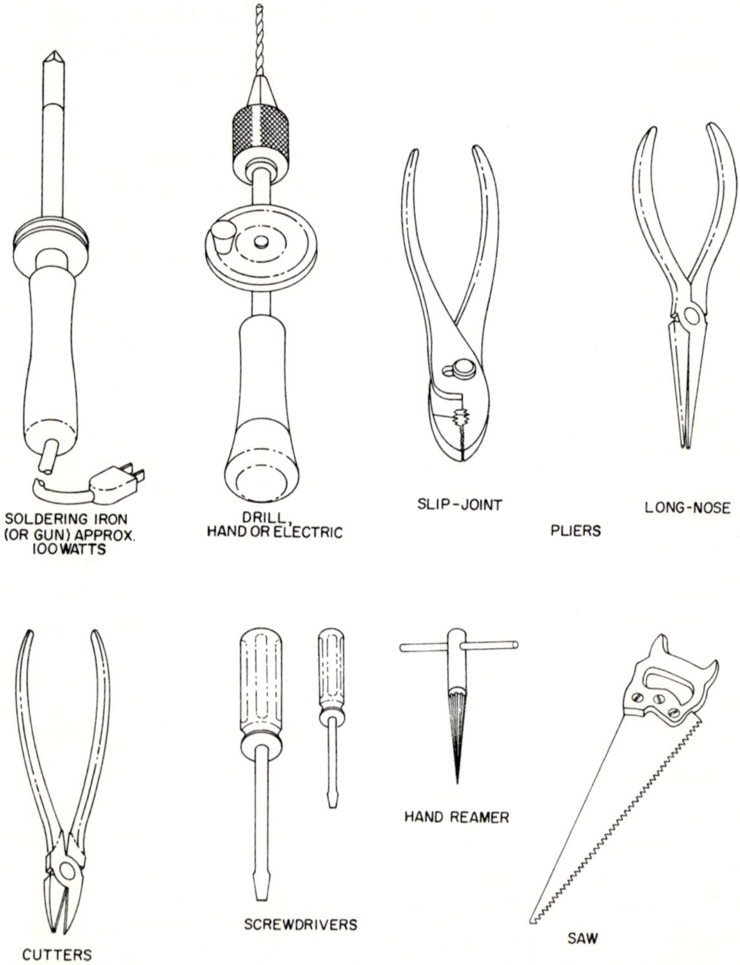

All of the games and toys in this book can be constructed with just the simple tools shown here.

TOOLS

Little in the way of tools is required. The basic complement should include a soldering iron or gun of approximately 100 watts, pliers, screwdrivers, and wire cutters. A drill (either the "egg beater" or electric type) and a saw will also be needed. The work will go more smoothly if you have a file and hand reamer for enlarging holes.

Be sure to use rosin-core solder only, which is made specifically for radio and TV use. The acid-core solder used in sheet-metal work will corrode the leads. A 60% tin–40% lead solder is recommended—just ask for "60/40."

No electronic test equipment is needed to build or adjust any project. However, in one or two places an ohmmeter is suggested for determining whether the connections on a part (usually a coil) are correct. If you don't own one, perhaps a radio amateur or other person interested in electronics will check the part for you.

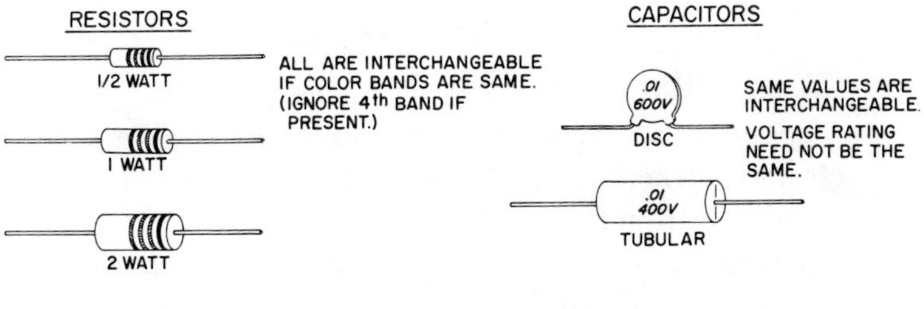

Parts do not have to be exactly the same as those specified. For example, ½-watt resistors are specified, but larger wattages will work just as well. Likewise, capacitors may be any type as long as the value is correct and the voltage rating is sufficient.

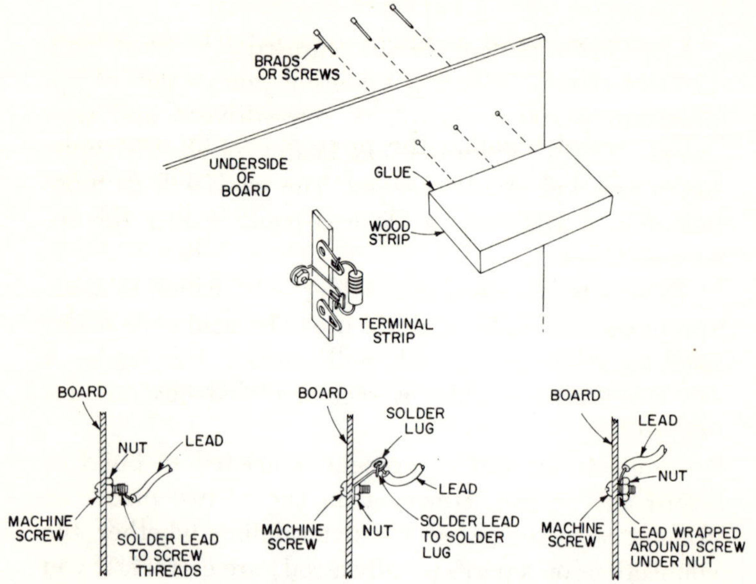

Mounting components to the hardboard. Wires may be attached to screws in any of the three ways. Glue wood strips to board and fasten with brads or screws.

PARTS

A parts list accompanies each chapter. Most of the components are standard types used in radio and TV sets and stocked by electronic parts distributors (or can be ordered from them). A certain leeway in their values is permitted. In fact, a departure of 20% either way should not noticeably affect circuit operation. Distributors are often able to recommend substitutes that will work as well as the parts listed.

Since certain parts reappear in several projects, you can save money by using them again. Unless you wish to keep the unit, don't cut the leads too short when installing components. You may not have enough lead left to re-use after it has been unsoldered.

A hardware store or lumber company is the source for composition board, wood strips, dowels, copper sheet (known as flashing), and other nonelectronic parts.

The cost of building the projects usually runs only a few dollars, rarely over five. The expenditure will be kept down if parts are handled carefully so they can be used again. The electronic hobbyist shouldn't overlook his "junk box," which often contains a wealth of serviceable components. Old radio or TV sets, too, can yield valuable parts.

There is considerable latitude in the choice of resistors. Although ½-watt units are called for, 1- or 2-watters are perfectly all right. Just match resistance values; the physical size is not too important. The same holds true for capacitors. Unless stated otherwise, the DC working voltage should be about twice the battery voltage. However, standard 400- and 600-volt capacitors will work and are usually less costly than low-voltage types.

Keep ease of handling in mind when picking hookup wire. Lamp cord will work, but a No. 20 or No. 22 plastic-covered type is more convenient. Use a tinned version for easier soldering. If the leads must undergo any flexing, they should be stranded.

CONSTRUCTION

A quick glance through the various chapters will reveal that most of the projects are built on a flat board. It may be hardboard (*Masonite*), composition board, plywood, or any other substance that can be easily sawed and drilled. By avoiding the use of metal, you will save much time and effort. The project will be more

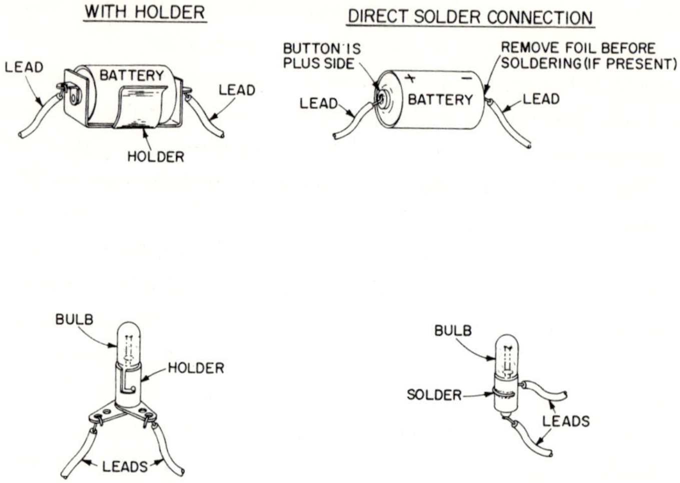

Batteries and bulbs may be mounted with holders, or by soldering directly to the components.

durable if both glue and small wire brads are used to join the wood supports to the board.

The parts and wires in the various illustrations are suggested layouts. You might want to follow them exactly, but there's ample room for variation. The circuits are not critical—i.e., their proper operation does not depend on very short wires or specific parts placement.

Where transistors are used, several possible type numbers are suggested in the Parts List. These may be of varying sizes and shapes, but all are designed to perform the same function. Be sure to refer to the terminal guide furnished with each unit when you connect the transistor in the circuit. The wiring diagrams in this book show one of the possible lead arrangements, but it may not be correct for the unit you are using.

In comparing construction techniques, you will occasionally find that a particular job is done one way in one project and a different way in another project.

Bulbs and batteries, for example, might be mounted in a holder in one unit, but have wires soldered directly to their terminals in another. Either method will work. If you are using the direct-solder system, apply the iron quickly to prevent excess heat from damaging the part. Batteries can be glued in place with *Duco* or similar cement if desired.

Check and recheck connections! A finished project that doesn't work is indeed a disappointment. To avoid errors, many hobbyists check off each connection on the diagram in pencil as it is soldered into the circuit, and also each part in the Parts List after it is installed, to be certain none have been overlooked.

When you are satisfied that your work agrees with the diagram, you may apply power—but not until then.

Have fun!

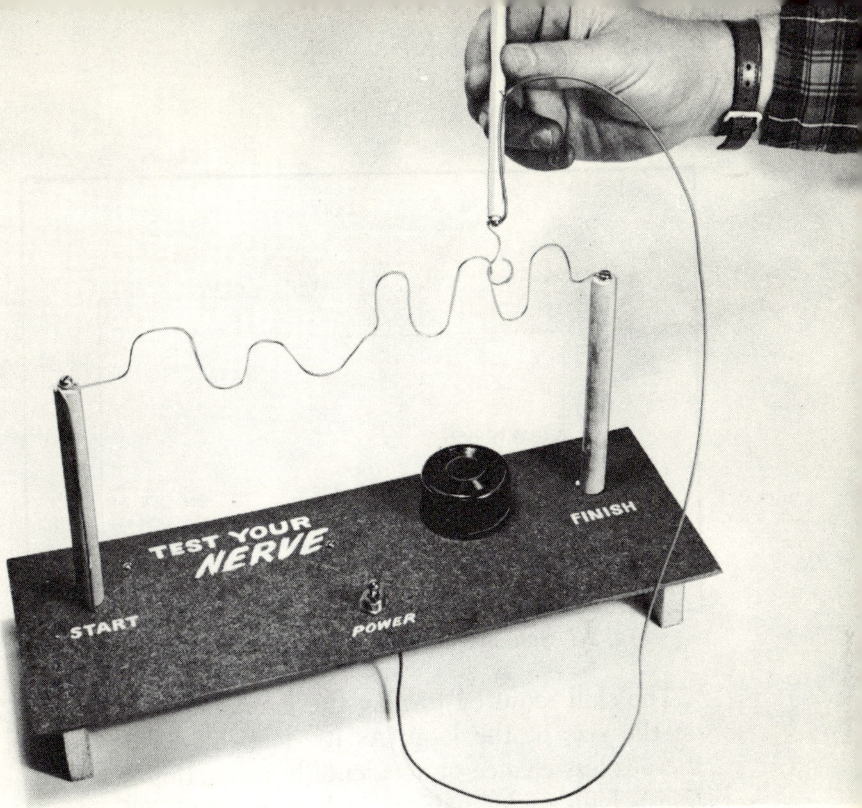

2 TEST YOUR NERVE

Although extremely simple in appearance and construction, "Test Your Nerve" puts your manual skill to a lively test. As shown in the above photograph, a small loop is hooked around the wire connected between the posts marked "Start" and "Finish." The object is to guide the loop over the weaving course *without touching the wire*. A slight misjudgment or quiver of the hand and the loop will contact the wire. This sets off a loud "foul" signal from the buzzer located just to the left of the "Finish" post.

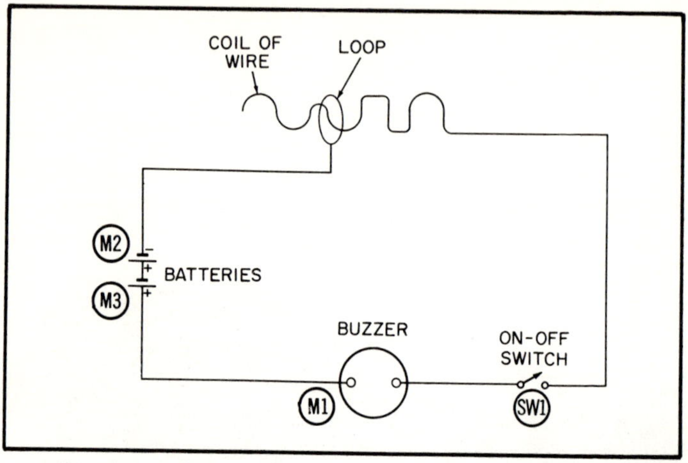

Fig. 2-1. Schematic of the "Test Your Nerve" game.

The skill required to play the game depends largely on the size of the loop. As its diameter is increased, there is less chance of accidentally contacting the wire. Use a loop under half an inch and you will frustrate the attempts of nearly every player. However, it is easy to enlarge the loop by bending it slightly. Another variable is the degree of twist and turn in the wire between the posts.

Scoring is a matter of counting the number of times the buzzer sounds during a single play. The person with the lowest total wins. If you find that the buzzer rarely sounds, make the loop an eighth of an inch smaller.

HOW IT WORKS

The schematic in Fig. 2-1 reveals a series arrangement of each circuit component. Whenever the loop touches the wire, the circuit is completed and the buzzer is energized. The power switch permits you to disable the circuit so the loop may be hooked onto the wire

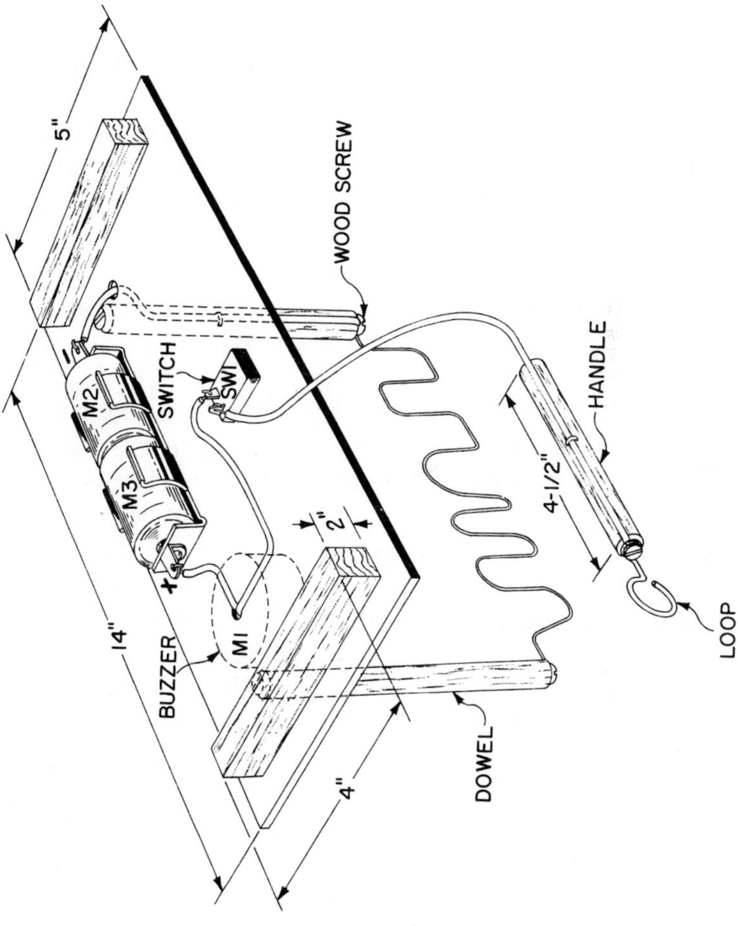

Fig. 2-2. Pictorial diagram of the game.

and the game stored away without the buzzer sounding continuously.

CONSTRUCTION

There is nothing critical in the assembly of this game. Start by cutting the baseboard to the dimensions shown in Fig. 2-2 from a piece of hardboard such as *Masonite*. Add the legs and you are ready to mount the two posts. Use half-inch wood dowels for them. Insert wood screws through the bottom of the board to hold the dowels in place. For added durability, glue all joints before tightening the screws. There is no preferred order for mounting the rest of the components—you may follow the illustrations if you wish.

No. 16 tinned copper wire is recommended between the two posts; it is readily bent into any pattern and will remain in any fixed position. The wire between the handle and the underside of the board should be flexible enough so it doesn't hinder movement of the loop between the posts. No. 20 stranded wire is a good choice. Fasten this wire to the handle (½-inch dowel) and to the wire loop with a wood screw in the end of the dowel. Attach the stranded wire to the handle with brads. If you followed the drawing in Fig. 2-2 fairly closely, your project should begin to look something like the one in Fig. 2-3.

Two standard flashlight cells are placed in the holder, providing a total supply of 3 volts. The buzzer (the kind used in code-practice oscillators) is rated at 1.5 volts. Although 3 volts may appear high for the buzzer, it will produce a sharp, positive action, even from the slightest brush of the loop against the wire. Note that the positive (button-side) terminal of the battery is to the left. Although the polarity of the batteries is not critical in

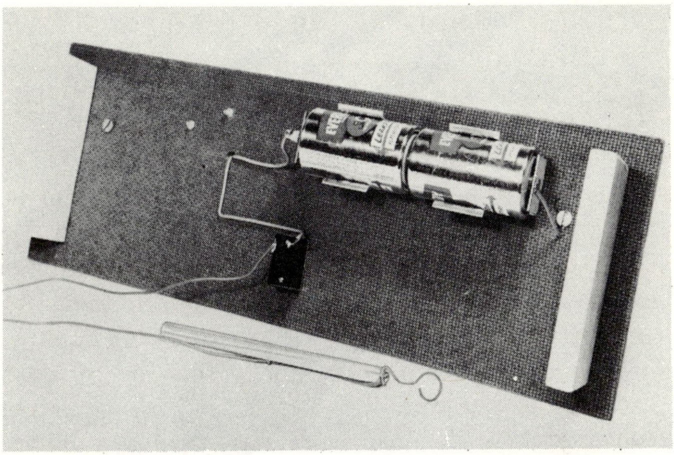

Fig. 2-3. Bottom view of the completed game.

this circuit, it is a good idea to form the habit of connecting the positive and negative terminals exactly as shown by the illustrations. This is extremely important with later projects. Otherwise, they may not work and components may even be damaged.

PARTS LIST

Item No.	*Quantity*	*Description*
SW1	1	Single-pole, single-throw toggle switch.
M1	1	1.5-volt buzzer.
M2, M3	2	Size D flashlight cells.
	1	5″ × 14″ piece of hardboard.
	15″	No. 16 tinned copper wire.
	24″	No. 20 stranded hookup wire.
	3	½″ wood dowels.
	2	⅞″ × 2″ × 4″ wood strips.
	1	Battery holder (for two size D cells).
		Wood screws.
		Brads.

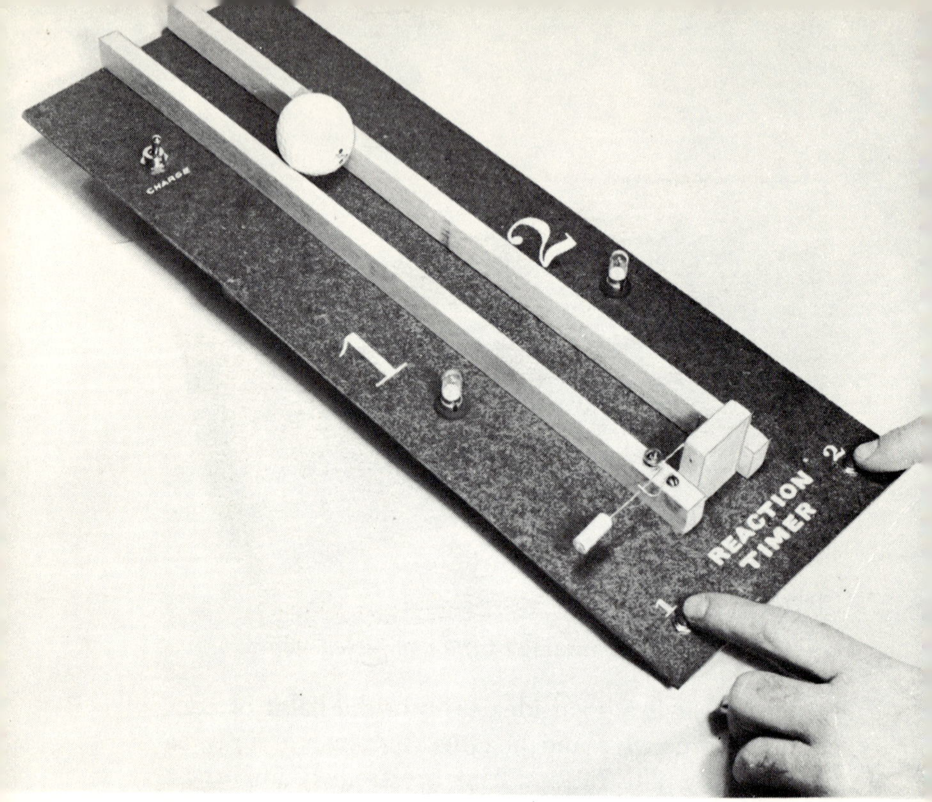

3. REACTION TIMER

Can you "split a second" more quickly than your opponent? You can find out with the "Reaction Timer," an electronic game that decides which player has the faster trigger finger. The basic setup is shown above. An ordinary golf ball is placed on the rails, at the top of the board. As it rolls down the slight incline, the two players stand poised, ready to press their respective buttons at the bottom of the board. The first player to press his button *after* the ball has struck the stop point is the winner—or the player with the faster reaction time. Note the large numbers *1* and *2* near the center of the

Fig. 3-1. Detail of the lower half of the board, showing ball about to strike pin of switch SW3.

board, just above the two bulbs. These correspond to similar numbers over the two player buttons. When a button is pressed, after the ball has reached the stop, the bulb on that player's side of the board will flash, signifying which one is the winner.

Several features, detailed in the next section, are incorporated to prevent an overanxious player from winning as a result of pressing his button before the ball reaches the bottom. A specially constructed switch, shown in Fig. 3-1, will cause both lights to flash indicating a foul if either player presses his button before the ball trips the trigger wire in front of the stop.

HOW IT WORKS

The heart of the circuit is a battery-capacitor combination. The schematic in Fig. 3-2 shows the electrical hookup. When SW4 is thrown to the right as shown, battery M3 charges capacitor C1. SW4 is the switch

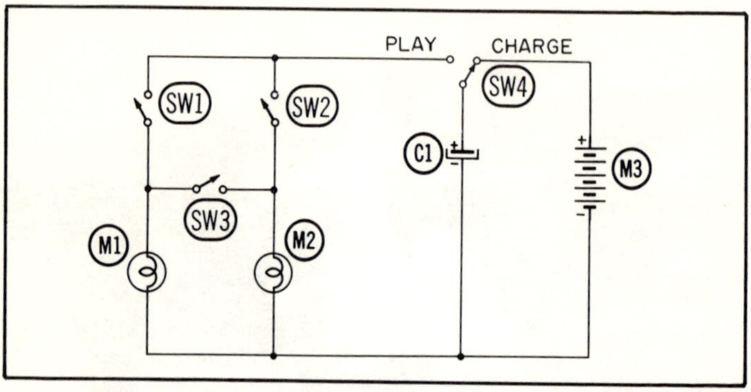

Fig. 3-2. Schematic of the "Reaction Timer."

(marked "Charge") at the upper left side of the board. Before each play it must be flipped to the charge position for a moment. This "arms" the game by allowing the capacitor to charge. When SW4 is in the play position, buttons SW1 and SW2 are ready to divert the charge through the two bulbs.

The rolling ball operates SW3. This switch remains closed and thus will cause both bulbs to glow dimly if either player presses his button before the ball strikes SW3. This constitutes a foul. When a foul is committed, the "Charge" switch must be momentarily flipped up and down and SW3 replaced in its closed position be-

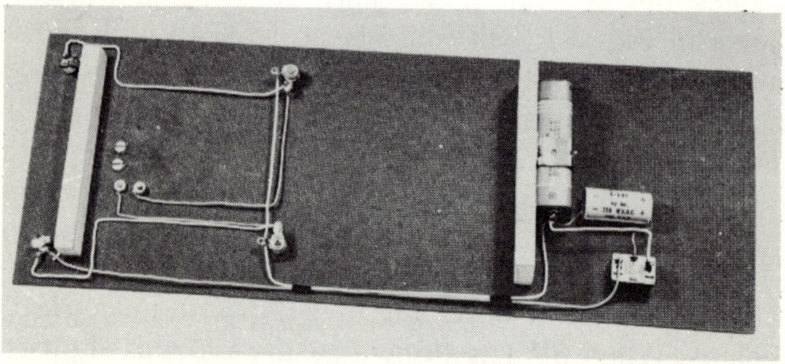

Fig. 3-3. Bottom view of the completed game.

Fig. 3-4. Detail of the battery mounting.

fore resuming play. However, when no foul has been committed, a bright flash by a single bulb indicates the player with the faster reaction time.

CONSTRUCTION

The underside of the board appears in Fig. 3-3. It is a piece of hardboard with two pine strips acting as the legs. Drill the holes to accommodate the switches and screws. Also drill two holes for the bulbs, and slip the rubber grommets into the openings. The bulb sockets can now be pressure-fitted into the grommets. The type suggested in the Parts List will comfortably hold the bulb sockets through friction alone.

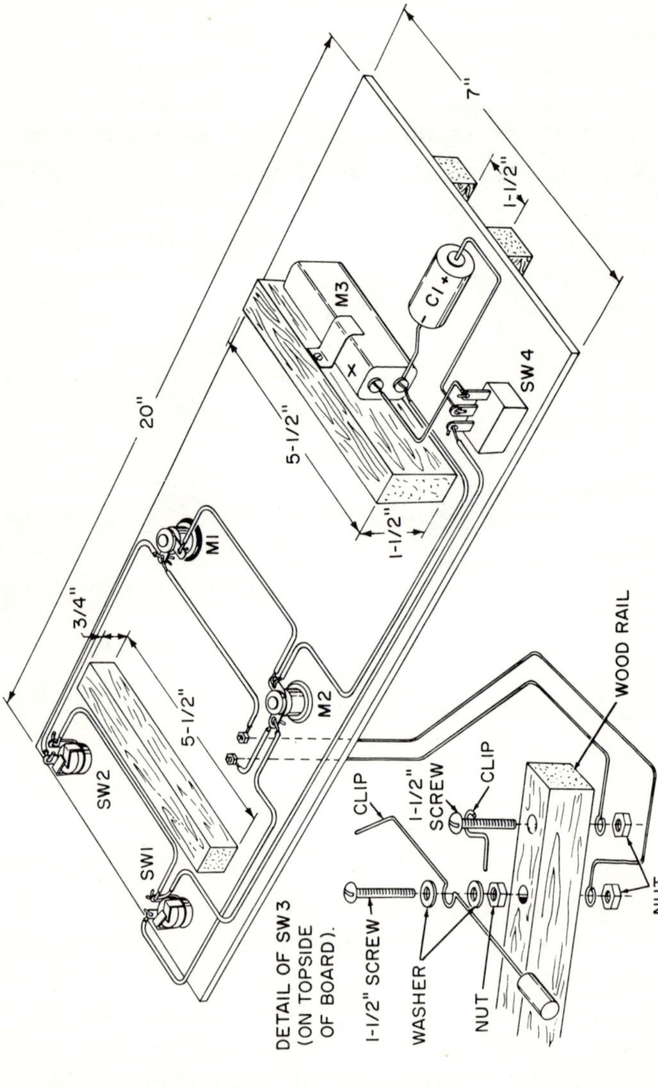

Fig. 3-5. Pictorial diagram of the "Reaction Timer."

The close-up in Fig. 3-4 shows an important aspect of the assembly—correct polarity of both battery and capacitor. The positive battery terminal is mounted *away from* the bottom of the board, and the positive lead of the capacitor connects to the center lug of switch SW4. CAUTION: An error in this connection will very likely ruin the capacitor. Note, too, the means of strapping the battery in place—a piece of scrap metal is bent into the shape shown and attached to the rear leg of the board. The battery leads in the example are simply soldered directly to the battery terminals. However, you may wish to use standard battery holders with tabs for making these connections. The battery should last an extremely long time. Further construction details are given in Fig. 3-5.

SW3, the ball-operated switch on the top side of the board, is made from a standard paper clip. The long top piece must swing freely after being mounted. Bend the short bottom piece so it presses firmly against the top piece when the switch is closed. (Check Fig. 3-1 for the approximate layout.) The wood dowel attached to the top piece is not intended primarily as a switch handle. Rather, it provides a reference point for the players to fix their eyes on. The instant it swings up, this signals the players that their buttons are "alive," ready to be pressed.

Leave the "charge" switch in the "play" position when the game is not in use. This will lengthen the life of the capacitor and the battery by removing the continuous battery charge. If a spring-return switch is used for SW4, it can be connected so it will automatically return to the "play" position. Thus, it can be pushed to the charge position and held for a moment and released. This way, it can never accidently be left on charge when not in use.

PARTS LIST

Item No.	*Quantity*	*Description*
C1	1	40-mfd, 150-volt DC, electrolytic capacitor.
SW1, SW2	2	Push-button switches (Argonne AR-375 or equiv.).
SW3	1	Switch constructed from paper clip (see text).
SW4	1	Single-pole, double-throw toggle switch.
M1, M2	2	No. 47 pilot lamps.
M3	1	67½-volt portable radio battery.
	2	Lamp sockets (for No. 47 bulbs).
	2	⅝" O.D. rubber grommets.
	1	7" × 20" piece hardboard.
	1	⅞" × 3¼" × 5½" wood strip.
	1	⅞" × 1½" × 5½" wood strip.
	2	⅞" × ⅞" × 17" wood strips.
		Machine screws with nuts and washers.
		Wood screws.
		Hookup wire.

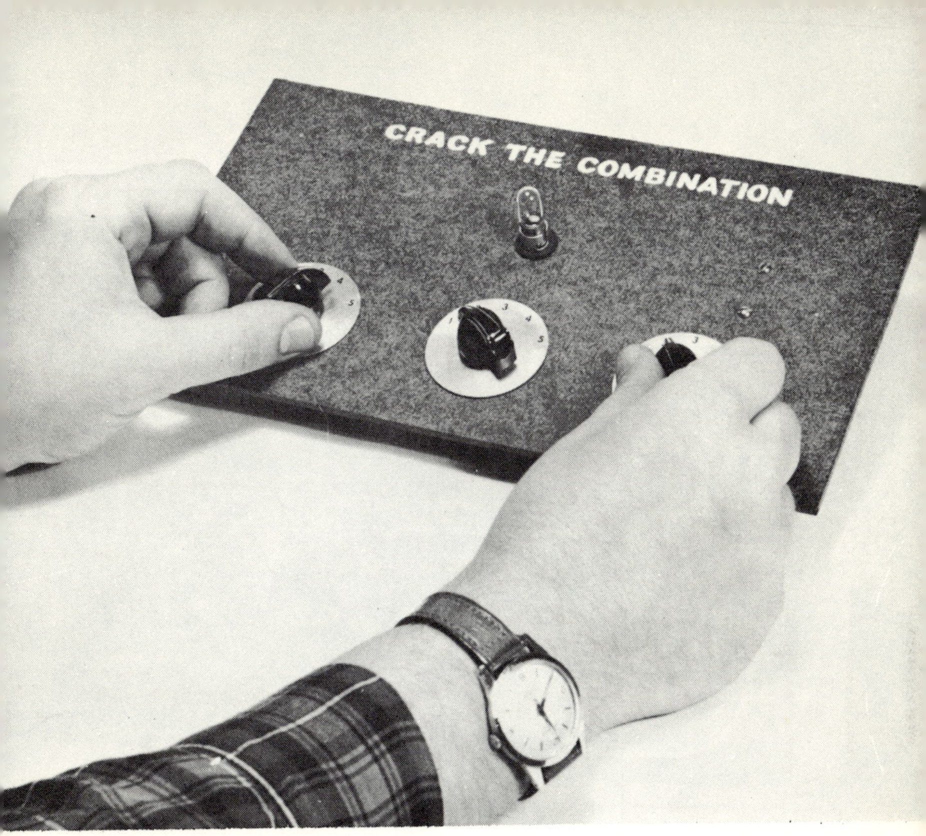

4. CRACK THE COMBINATION

The talent of a safecracker plus a sprinkling of logic are needed to "Crack the Combination." The game can be readily adapted to challenge the skill of a young child or professional mathematician, depending on how it is set up in advance.

A look at the above photograph reveals three knobs along the bottom edge of the board. Each can be rotated to any of five positions. The player must manipu-

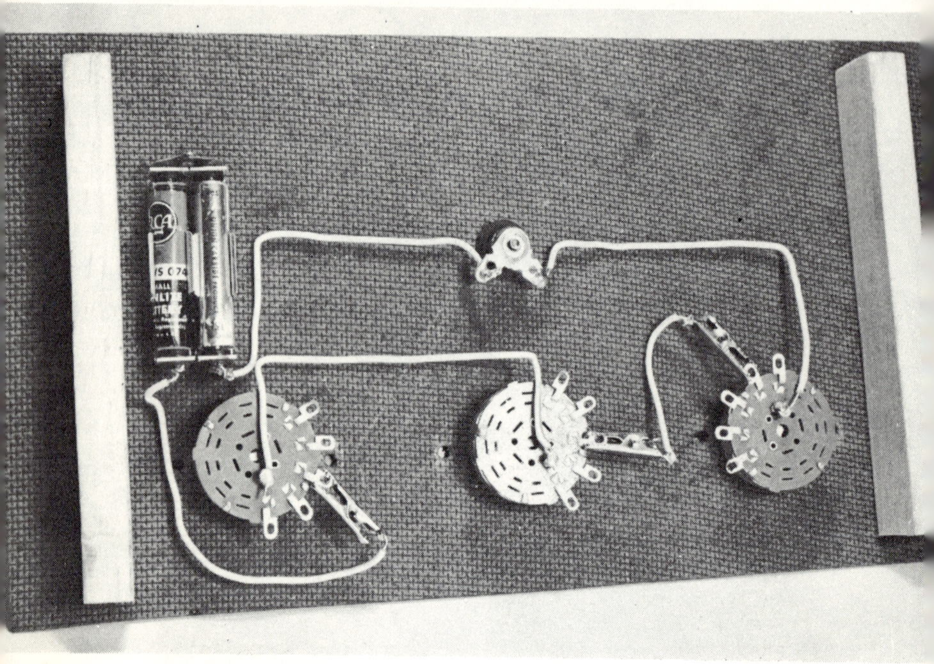

Fig. 4-1. Bottom view of the board. Batteries are at the left and switches are along the bottom.

late the knobs in as many combinations as possible. When he arrives at the correct one (preset in advance), the small lamp near the center of the board lights and scores a win. The number of possible combinations runs as high as 125 in the most difficult version of the game. However, the level of difficulty can be readily changed, depending on the method of setting up the combination.

Fig. 4-1 shows the underside of the board. Note the rear of the three switches; each has five lugs in a semicircle. These are the contact points, which correspond to the numbered knob positions on the front of the board. The three miniature alligator clips enable the combination to be changed by merely shifting them

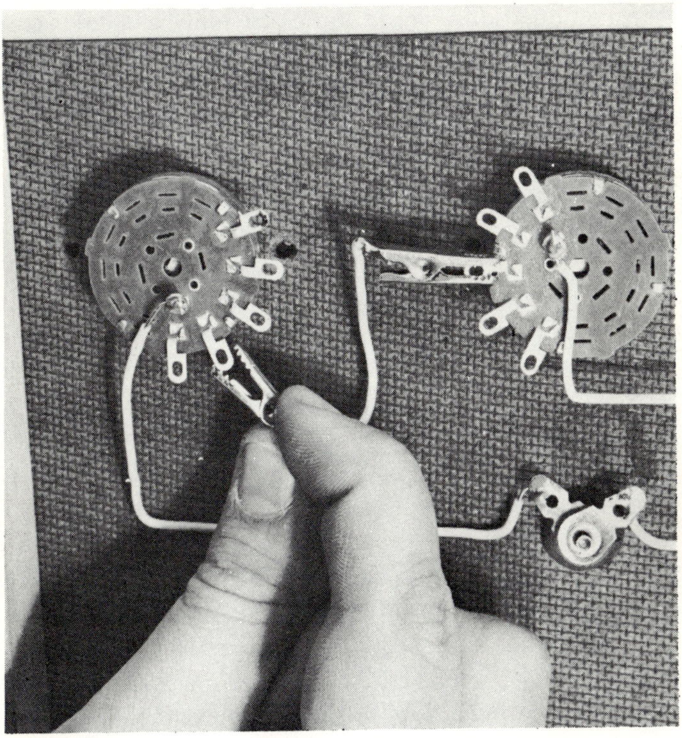

Fig. 4-2. Changing the combination.

among the various lugs. A player can thus pit his skill against the game repeatedly. The hand in Fig. 4-2 is shown moving a clip from lug 2 of the first switch.

Aside from changing the combination, the clips determine the game's level of difficulty. For youngsters, you can use only the first three positions on each switch. By thus avoiding lugs 4 and 5, the possible combinations are reduced to 27. For adults, especially persons who are mathematically inclined, use the full complement of five per switch.

Some players may demand the toughest version, but soon become confounded by their inability to "crack the combination." Some judicious hints can take care of this situation. If a player is making only slight progress, drop such hints as "Number 5 is not used on the first switch," etc. You, of course, know the combination since you set it up beforehand.

HOW IT WORKS

The schematic of the game appears in Fig. 4-3. Bulb M1 and batteries M2 and M3 are connected in series with each other and the circuit is completed through the switches. J1, the alligator clip from the negative battery terminal, is clipped to any of the five numbered lugs on the first switch, SW1. In the position shown in Fig. 4-3, when the arm of the switch is rotated to lug 1, battery voltage will proceed to the next switch, SW2. Alligator clip J2 picks it up if the player rotates this switch arm so it comes to rest on lug 4. The last switch, SW3, completes the circuit to the bulb when its arm contacts lug 5. Thus, the combination in this example is 1-4-5. The bulb does not glow unless the player sets up this particular switch arrangement.

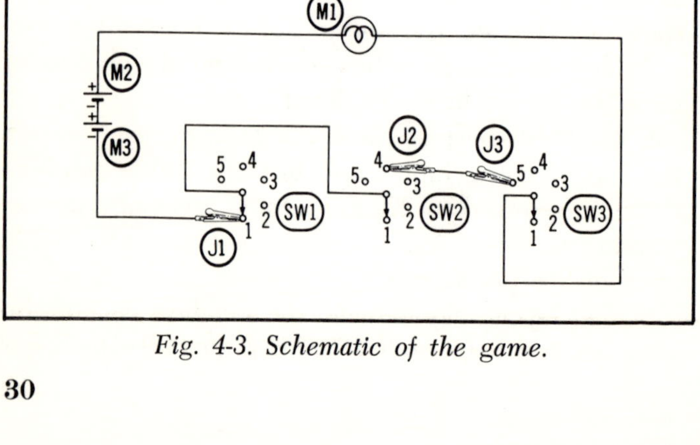

Fig. 4-3. Schematic of the game.

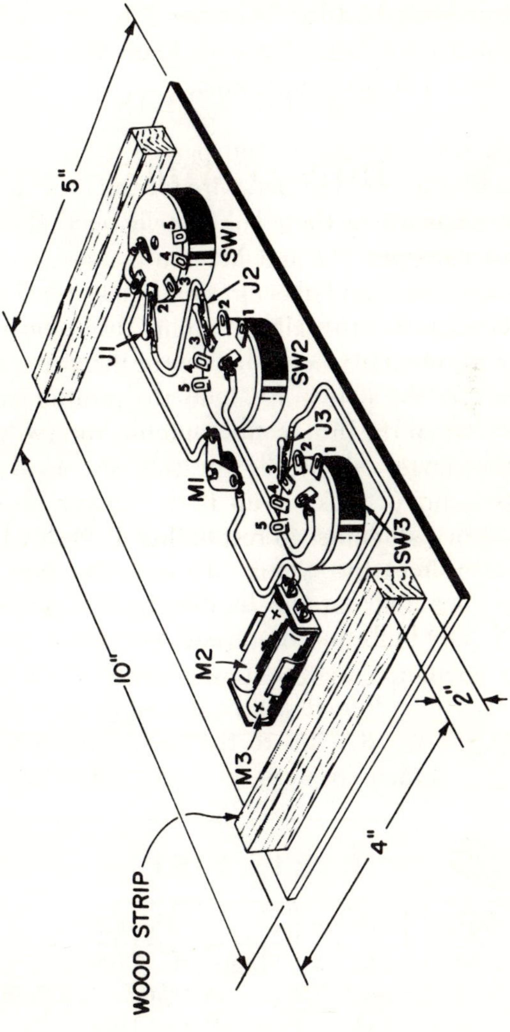

Fig. 4-4. Pictorial diagram of the game.

As mentioned earlier, the game is least challenging when fewer than four or five contacts on each switch are selected by the alligator clips.

CONSTRUCTION

Follow the illustrations and parts list for the assembly. It is best to use stranded wire for the leads to which alligator clips are conected. You will need their flexibility for changing the combination.

The switches may be any single-pole, five-position rotary type. Just be sure they are of the "nonshorting" kind, which prevents more than one contact from being energized as the knobs are rotated. Viewed from the rear, or lug side, the numbering of the switches is counterclockwise. This is illustrated in the wiring guide of Fig. 4-4.

The switch dials may be purchased, or made from any suitable metal or plastic. In fact, the points may simply be painted on the hardboard if desired.

Drill a hole for lamp M1 and fit the rubber grommet into the opening. Then insert the lamp socket into the grommet. The grommet will hold it in place.

Mount the wood strips to the hardboard with wood screws. For added strength, glue the strips first.

PARTS LIST

Item No.	*Quantity*	*Description*
SW1, SW2, SW3	3	Single-pole 5-position, nonshorting rotary switches.
M1	1	No. 49 pilot lamp.
M2, M3	2	Size A flashlight cells.
J1, J2, J3	3	Alligator clips.

(continued)

PARTS LIST (cont'd)

Item No.	*Quantity*	*Description*
	1	Lamp socket (for No. 49 lamp).
	1	Battery holder (for two size A cells).
	3	Switch dials (numbered 1 to 5).
	1	⅝″ O.D. rubber grommet.
	1	Piece hardboard (approximately 5″ × 10″).
	2	⅞″ × 2″ × 4″ wood strips.
		Stranded hookup wire.
		Solid hookup wire.
		Wood screws.

5. PENNY PITCH

The price of admission to this game is merely a few pennies. "Penny Pitch" can be played anywhere—on a floor, desk, or table—and it provides a challenge to the players' pitching ability. An over-all view of the project is given above. Five bull's-eyes make up the targets. If a penny touches both the center and the outer rim of any bull's-eye, the bell near the top of the board will sound off. When any target *except* the middle one is

Fig. 5-1. Close-up view of the center target.

struck, the score is five points. The middle target deserves just a single point; it is much easier to hit than the other four near the edges. The close-up in Fig. 5-1 shows why. If the penny rests on the circle plus any one of the five other contact points, the bell will ring. Other targets have just one contact point, in the center of each circle. This allows great opportunity for scoring on the middle one, and for this reason contributes less to the total score.

To set up the game, place the board about five feet away from the pitching position. It is best to mount it somewhat lower than the player's hand. A higher position results in an unnatural swing of the arm and makes scoring far more difficult. Of course, the distance can be changed according to the players' skill.

HOW IT WORKS

The circuit consists of five switches which correspond to the targets, as shown in the schematic of Fig. 5-2. The black dots of each switch represent the center contacts. For example, when a penny lands across the black

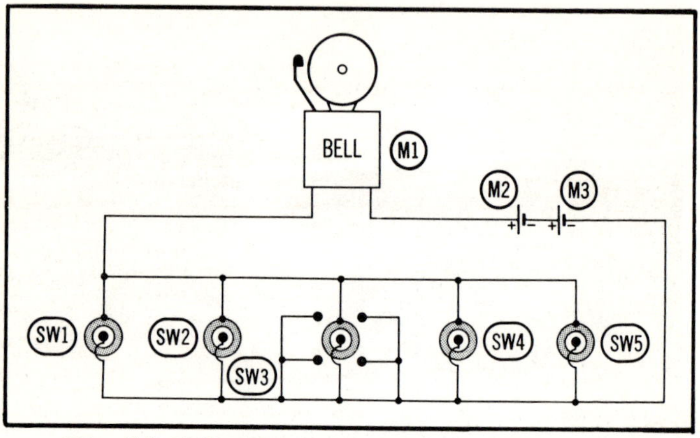

Fig. 5-2. Schematic of the "Penny Pitch" game.

dot and the circle around SW1, the circuit from bell to battery is completed and the bell rings. SW3, the middle target, has four extra points connected to its center contact. No on-off switch is required, since the battery is not energized except when a penny completes the circuit.

CONSTRUCTION

Cut the board to 17 by 20 inches. Drill a small hole for the bell wires, and mount the bell. Next mount the battery holder, as shown in Fig. 5-3. It does not matter which way the batteries are inserted into the holder, as long as they are in series (plus to minus, or facing in the same direction).

The construction of the targets is detailed in Fig. 5-4. Cut five circles of copper from ordinary copper flashing (obtainable at many hardware stores). Outside diameter is 1½ inch, and inside diameter is ⅝ inch. A file is handy for smoothing the rough edges. Flatten the circles with a hammer, and shine them with steel wool.

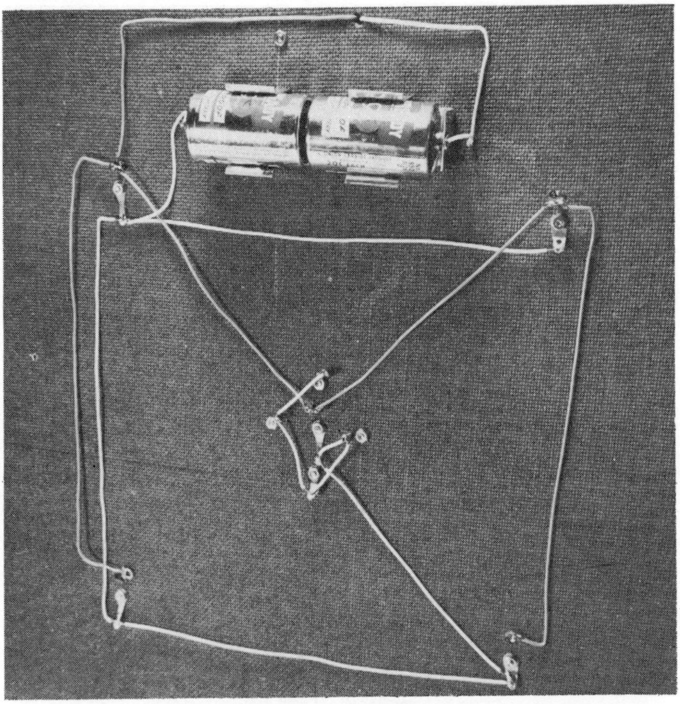

Fig. 5-3. Bottom view of the completed game.

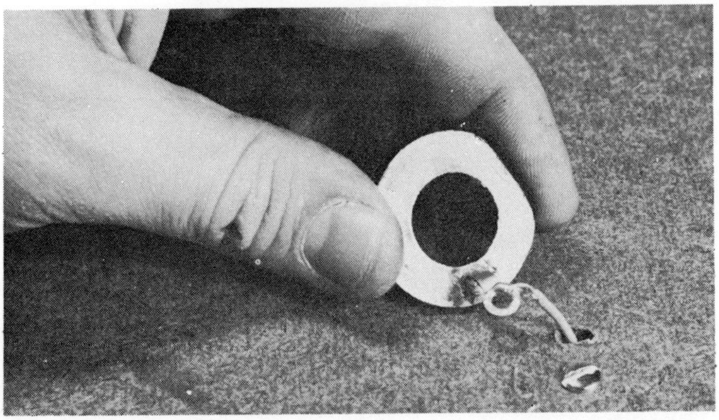

Fig. 5-4. Attaching copper ring to the board. Prepare as shown and glue to board.

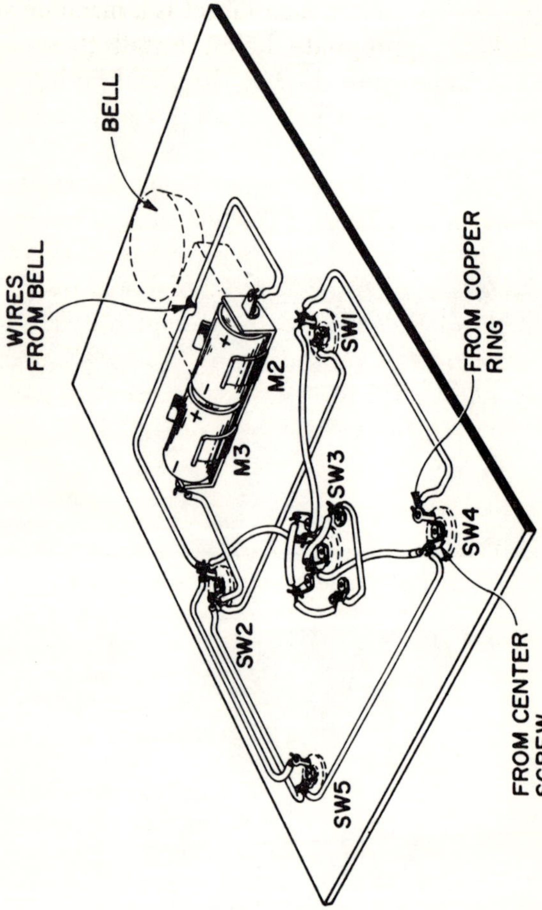

Fig. 5-5. Pictorial diagram of the "Penny Pitch" game.

You can solder a wire directly to each one, or use a solder lug as shown in the model. Install the targets on the board by drilling a hole for the wire and gluing each copper circle in place.

The center contact of each target is a machine screw. After drilling appropriate holes, install these screws with a nut underneath the board to hold both the wire and screw. Solder lugs under each nut will simplify the wire connections.

Fig. 5-5 is the wiring guide. Don't insert the two batteries until last. Then, if the bell rings when they are inserted, you know a short circuit exists. If there is no error, any metal across the contacts should operate the bell. There is only one precaution—look out for pennies which are extremely dirty or greasy. They will not make good electrical contact. A few swipes of steel wool will quickly remove the interfering insulation.

PARTS LIST

Item No.	*Quantity*	*Description*
SW1 through SW5	5	Switches (construct as described in text).
M1	1	6-8–volt doorbell.
M2, M3	2	Size D flashlight cells.
	1	Battery holder (for 2 size D cells).
	1	17″ × 20″ piece of hardboard.
	9	6-32 machine screws and nuts.
	14	Solder lugs.
		Hookup wire.
		Copper flashing.

6. ELECTRONIC CARD TRICK

Without sleight of hand or cards up your sleeve, you can perform some entertaining wizardry with the "Electronic Card Trick." It baffles up to six people at a time as you correctly "guess" the two cards each person has selected at random from the deck. You, the magician, manipulate a mysterious switch near the bottom of the board while an electronic "brain" lights up its bulbs to indicate where cards are placed and which ones were

Fig. 6-1. Top view of the completed board. Cards will be placed over white squares as trick is played.

chosen. A detailed description of how to perform this trick appears at the end of the chapter.

HOW IT WORKS

The secret is in the wiring of a 12-point switch to a group of 12 pilot lamps. The switch is visible near the bottom of the board in Fig. 6-1. The corresponding lamps are positioned above the white squares painted on the board's surface. Rotating the switch instructs the players where to place the cards. The circuit pattern is calculated so that cards are distributed strategically in pairs that the switch "recalls." Further switch rotation, later in the trick, informs the operator of the players' choices.

Fig. 6-2. Close-up of the switch construction.

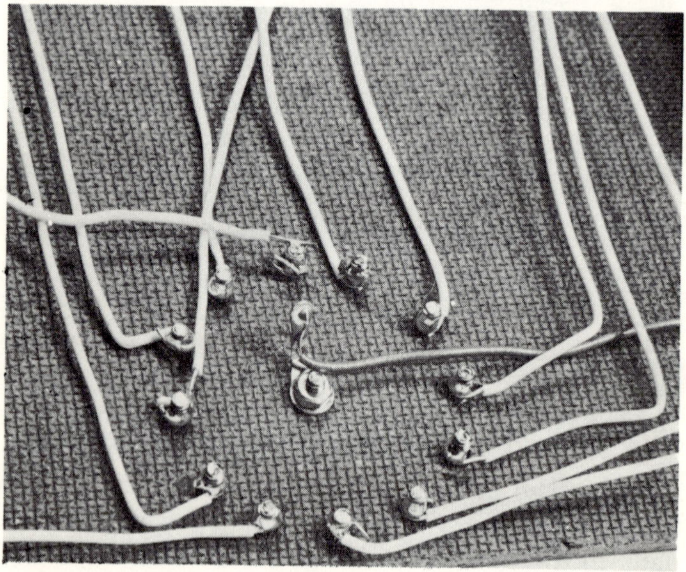

Fig. 6-3. Close-up view of the bottom of the switch.

CONSTRUCTION

Start with a piece of board about 12 by 18 inches. Paint twelve 2-inch squares on its surface, as shown in the model. The distance from the edge of one square to the next is ¾ inch from left to right, and 2¼ inch from the bottom of one to the top of the next. It's helpful to use twelve playing cards as a guide in marking the board. In this way, you can be certain to leave enough room at the top of each square for mounting the pilot lamp.

Next mount the lamps. By drilling holes just under ⅜ inch in diameter, you can force-fit the lamps into them and eliminate bulb sockets. However, if sockets are available they should be used since they will simplify the wiring, as explained later. If sockets are used, the holes will have to be larger to accommodate the socket. The switch at the bottom center of the board is a group of machine screws and nuts arranged in a circular pattern. Draw a circle two inches in diameter, and space the 12 screw holes equally around the circumference. The *centers* of the screws should be about half an inch apart. No. 6-32 screws were used in the model, but any convenient size will serve. Mount one screw at the center of the circle, and secure the switch lever under its head. The lever is a piece of copper (sold as "roof flashing" in hardware stores) cut to 1¾" × ½". Examine Fig. 6-2 for a close-up of how the switch should look on top of the board. Note that the free end of the lever is bent up into a tab which acts as a handle. Also, the two sides of the lever are bent upward slightly where they contact the screw heads. This allows the lever to be rotated freely over the screws in both directions. Be sure to bend the lever up and away from the screw heads so it does not make contact until pushed down. This prevents the bulbs from burning while the game is not in use.

Underneath the board (see Fig. 6-3), the switch is completed by installing a solder lug and nut to the center screw. If the screws are plated, they will take solder readily and you can solder wires directly to the threads of the other twelve. Otherwise, use nuts to secure the wiring.

The rest of the construction involves installing the battery and wiring between the lamps and switch. An over-all view of the board's underside appears in Fig. 6-4. The wires can be soldered directly to the battery and the battery cemented to the board as shown in the model; but if available, a battery holder is a much more desirable method of mounting.

An important note on batteries: Many manufacturers install a shiny foil covering on the negative (flat) end of their batteries. This is no problem with flashlights, where batteries are compressed by a spring, but the foil often prevents proper contact when wires are soldered to it. The remedy is to use a pointed instrument to pry away and remove the foil completely. With the foil removed, a secure solder connection can be made.

The wiring pattern for the unit appears in Fig. 6-5. If sockets are used for the lamps, take a bare wire and thread an end through the terminal on each socket which connects to the brass shell. Then run an insulated wire from each switch terminal to the center-contact terminal of the corresponding bulb socket. Solder both terminals on each socket. If sockets are not employed, run a bare wire to the brass shell of each pilot lamp. Form a tight loop about the shell, and fasten the wire in place with a touch of solder. Run an insulated wire from each switch terminal to the center contact of its associated lamp (see Fig. 6-5). Be extremely

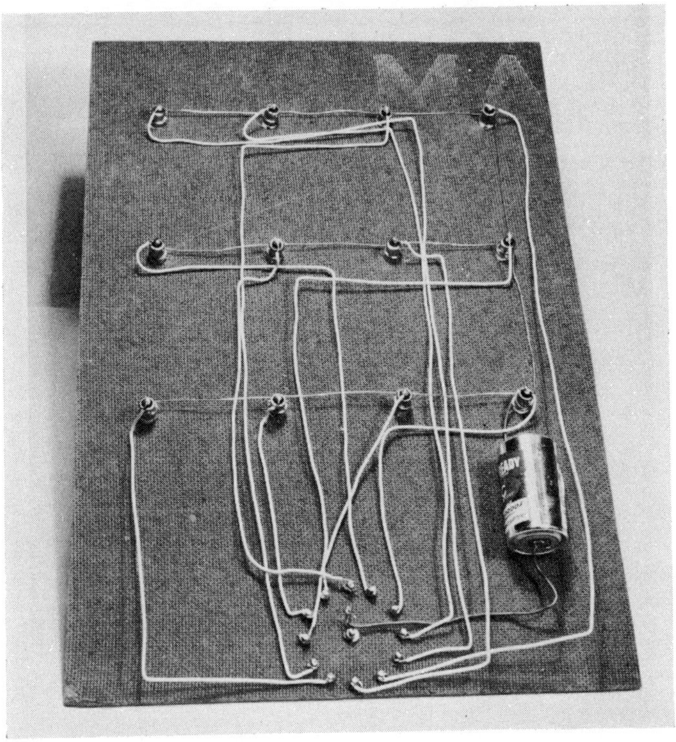

Fig. 6-4. Bottom view of the completed board.

careful when soldering to the center contacts of the bulbs—use just enough heat to melt the bubble of solder there.

The schematic of the unit appears in Fig. 6-6. Notice that the topmost screw of the switch is designated "Terminal 1." The rest are numbered counterclockwise for a total of 12.

PERFORMING THE TRICK

Each time you play the trick, point the switch lever directly toward the top of the board—poised over the

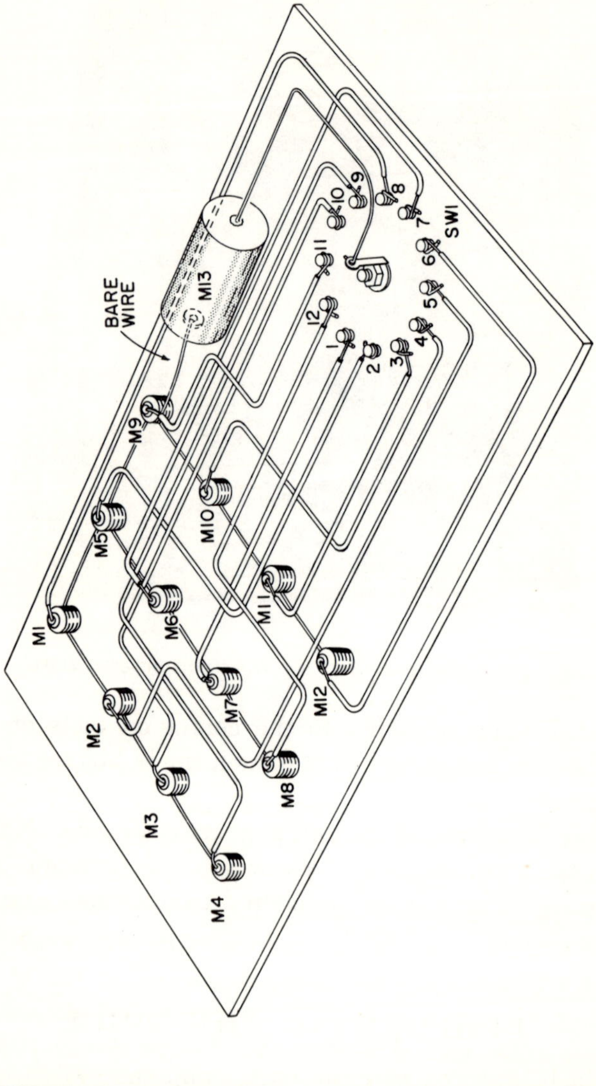

Fig. 6-5. Pictorial view of the "Electronic Card Trick."

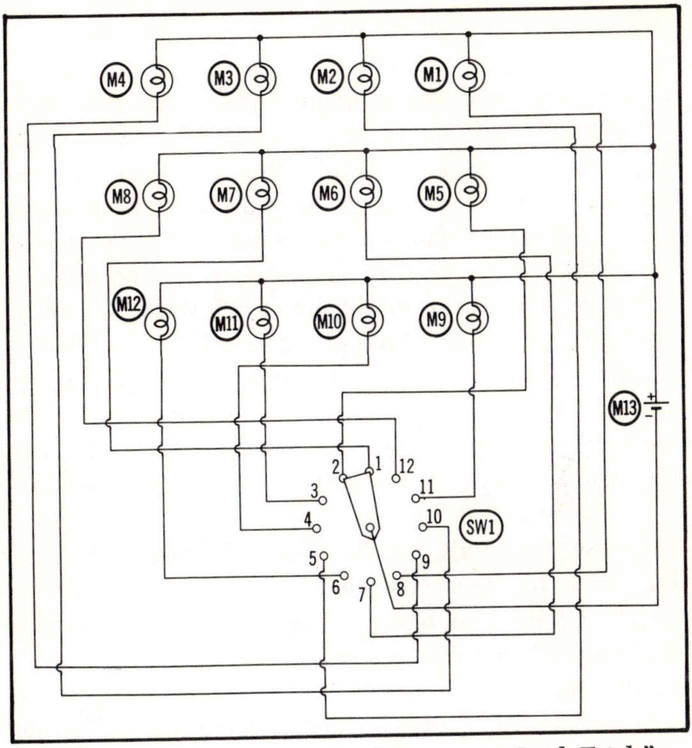

Fig. 6-6. Schematic of the "Electronic Card Trick."

screw corresponding to Terminal 1. Instruct the members of your "audience"—who may be any number from one to six—to shuffle a deck of cards to their complete satisfaction. Have one person deal off six piles of cards, two to a pile. Pairs are placed *face down* on the table; the "magician" must not see their faces. Tell each person to select one pile and memorize the two cards in it. After this is done, all piles are stacked on top of each other. Here you have to be cautious. The *piles* do not have to be stacked in any order, but be certain that the two cards in each pair remain together. Don't allow any further shuffling at this point.

Next have one of the players remove one card at a time from the top of the stack and place it wherever he sees a lighted bulb on the board. This calls for a bit of showmanship. Different lamps will light, seemingly at random, as you rotate the lever from Terminal 1 to 12. The switch can be moved one step at a time and a card placed on each square that lights. To prevent the operation of the trick from becoming too obvious, however, it is best to rotate the switch several times, so the lights will flash on and off all over the board, before asking the player to place the next card on the lighted square. The key is that each two successive stops of the switch rotation must be made at adjacent terminals on the switch. That is, if the switch is stopped at Terminal 1 for the first card, the next stop must be at terminal 2. Likewise, terminals 3-4, 5-6, 7-8, 9-10, and 11-12 must be paired together. The order in which you take the terminal pairs makes no difference. For example, you could stop at terminal 4, then 3, then 11 and 12, etc. A small white dot painted between the paired-up terminals will aid in identifying the terminal pairs. The dot is also useful for determining which cards the player has selected, as will be explained later.

After the cards have all been distributed, explain to the players that there are three rows of four cards each. Ask each player, in turn, to call out the row or rows that contain his cards. He will say, for example, "Top and bottom row," or simply "Middle row." Rotating the switch in a special way will now reveal the card pairs chosen by each player. Here is how.

Start by placing the lever *between* terminals 1 and 2, and pressing down. The first and third lamps of the middle row will light up as the lever contacts two screw heads simultaneously. If the player has called out the middle row, the cards below the two lighted bulbs are his. Find the other players' cards by pressing the lever between other terminal pairs—3 and 4, 5 and 6, 7 and 8, 9 and 10, 11 and 12. You can always tell the chosen pair by observing which lights are glowing—they will agree with the one or more rows called out by the player. Table 6-1 gives the rows and the bulbs which will light for each switch combination. When playing with less than six people, simply ignore the vacant squares.

After experimenting with the game for a while, you will find that all steps take only a minute or two. Of course, the operation of the switch is never explained to the players. All they see are bulbs going on and off, seemingly at random. The mystery is greatly increased if you don't pause when the lighted bulbs agree with the rows called out. Instead, keep on rotating the switch before announcing the card pairs. With just such a system, the model shown in the photographs has fooled more than one electronic engineer. Use some showmanship, too. As you rotate the switch, say that the electronic brain is "thinking," etc.

As mentioned before, it is helpful to paint a small white dot between each terminal pair. Then it will be an easy matter to rotate the switch lever without hesitation and find the correct positions for lighting two lamps at a time. Be careful not to get an incorrect read-

Table 6-1. Lighting Sequence of Bulbs

SW Terminals	*Rows*	*Lighted Bulbs*
1-2	Middle	M7-M5
3-4	Bottom	M11-M10
5-6	Top & Bottom	M2-M12
7-8	Middle & Top	M6-M1
9-10	Top	M4-M3
11-12	Bottom & Middle	M9-M8

ing by pressing down between terminals 2 and 3, 4 and 5, etc. With a certain amount of rehearsal, the trick will amaze and entertain while losing little of its "magic" through repetition.

PARTS LIST

Item No.	*Quantity*	*Description*
SW1	1	Switch (see text).
M1 through M12	12	No. 49 pilot lamps.
M13	1	Size-D flashlight cell.
	1	12″ × 18″ piece hardboard.
	13	6-32 machine screws and nuts.
	13	Solder lugs.
	1	1¾″ × ½″ copper strip (for switch).
		Hookup wire.

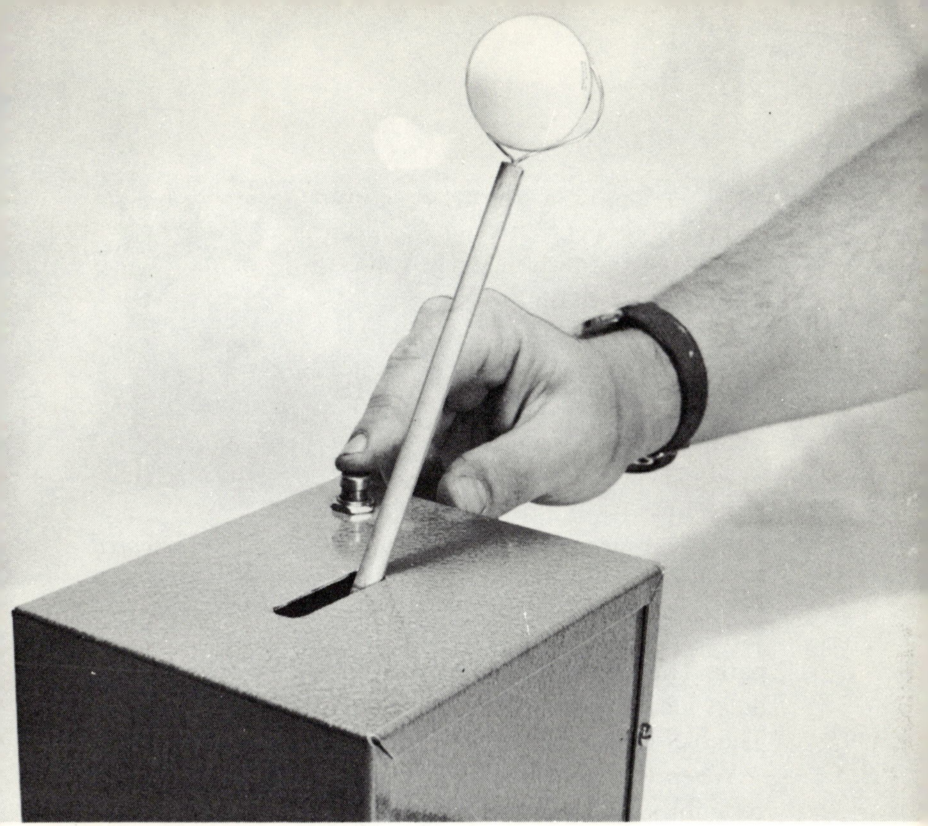

7. THE "FLIPPER"

A good eye and a measure of skill make the "Flipper" an interesting game. The setup appears above. A wooden arm supports a ping-pong ball as shown. After taking careful aim at the target, the player presses the button atop the metal case. The arm flips forward vigorously, shooting the ball many feet across the room to the target (shown in Fig. 7-1).

An examination of the target reveals two basic ways of scoring a hit. When the case is left in a horizontal position as illustrated, the ball will follow a fairly level

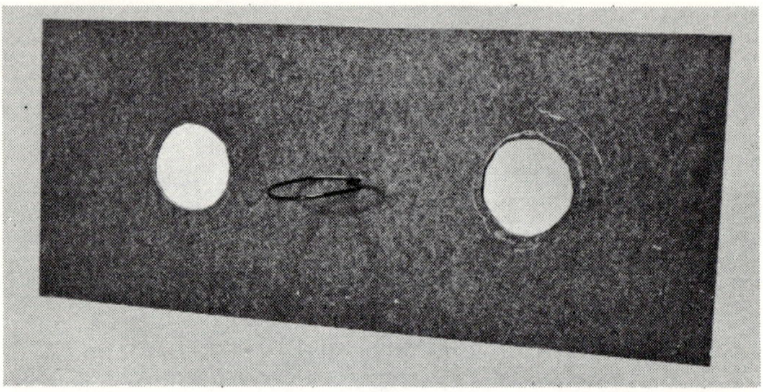

Fig. 7-1. Appearance of completed target. Hits are scored by shooting the ping-pong ball through the two holes. The hoop at the center allows you to "play basketball."

path. This position permits scoring in either of the two holes located at the left and right of the target center. The ball is aimed by shifting the case from side to side.

When the hoop between the holes is used, the game resembles the tossing of a basketball. Here the ball must follow an arc in order to drop into the hoop, and another element—trajectory—is introduced. The case is grasped in both hands, leaving the thumb free to press the button. When the case is tilted upward, the ball will describe a path like that of a ballistic missile. Depending on the player's skill, the ball will fall long or short according to the angle of tilt.

In working out a scoring system, assign three times as many points to a hit that makes the ball fall through the hoop. The added factor of trajectory makes the hoop shot more difficult.

HOW IT WORKS

A solenoid inside the case supplies the quick burst of energy needed to flip the ping-pong ball. The device

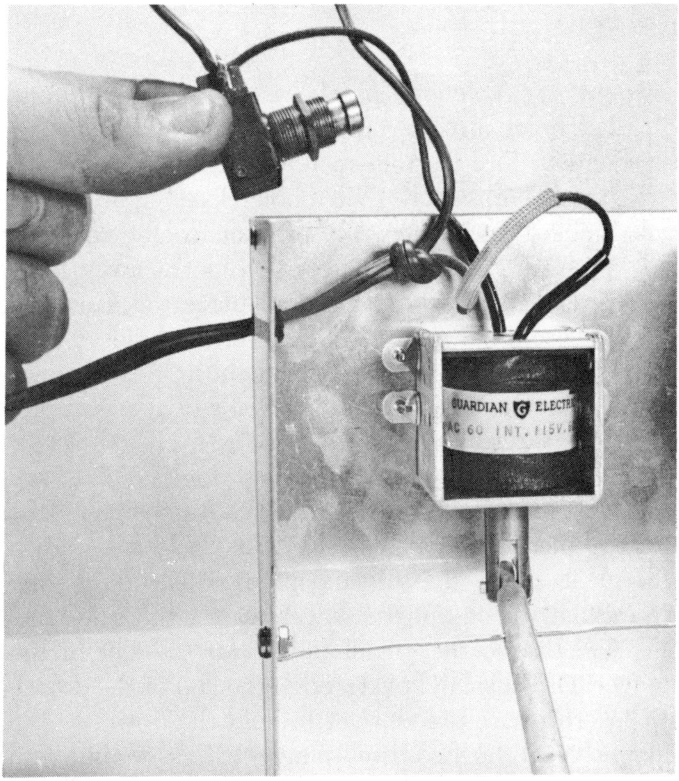

Fig. 7-2. Bottom half of case with cover removed to show solenoid, wooden arm, and wire axle. The hand holds the AC push-button switch.

consists of a coil and plunger. Each time the button is depressed, a surge of current is sent into the coil, magnetizing it. The plunger is pulled back abruptly, and a simple linkage drives the wooden arm forward. Fig. 7-2 shows the position of the solenoid within the case. Note that the wooden arm connects to the plunger and also rotates on an axle. Thus, as the plunger is pulled into the coil, the free end of the arm is driven in the opposite direction.

CONSTRUCTION

Mount the solenoid in the case, using the four threaded holes already tapped in the metal frame of the solenoid. Fashion four mounting brackets, each ¾″ × ⅜″, from scrap metal. Drill a hole close to one end of each bracket, and secure the brackets to the solenoid body with 6-32 screws. Bend each bracket away from the solenoid at a right angle, and place the assembly into the case. Four additional holes in the bracket ends accommodate nuts and bolts for mounting the solenoid. Center the solenoid between the walls of the box, one inch from the rear edge (see Fig. 7-2).

Cut the arm from a piece of wood dowel 8¾″ × ⅜″. Drill a hole 1½ inches from one end for the axle, and insert a piece of stiff wire (a coat hanger is ideal) through it. Using the solenoid plunger as a guide, file or sand flat the end of the dowel nearest the axle hole. This will enable the dowel to fit into the slot in the plunger. Drill a small hole through the end of the dowel, and insert a wire pin or rivet to hold the dowel to the plunger. Next, fit the dowel-plunger-axle assembly onto the walls of the case. Drill two holes into the sides of the case 3½ inches from the rear edge and 2 inches from the bottom. Using nuts and bolts, mount the ends of the wire axle to these bolts. (For a stronger mechanical support, loop the wire around the screw threads.) The assembly does not have to be absolutely rigid; in fact, some play is desirable to prevent it from binding during a game.

Drill a hole for the AC line cord. Then connect the cord, switch, and solenoid as shown in Fig. 7-3. Slip a piece of "spaghetti" (insulation) over the line cord, and solder the leads together. Slide the spaghetti back over the connections, and make sure it will not slip out of

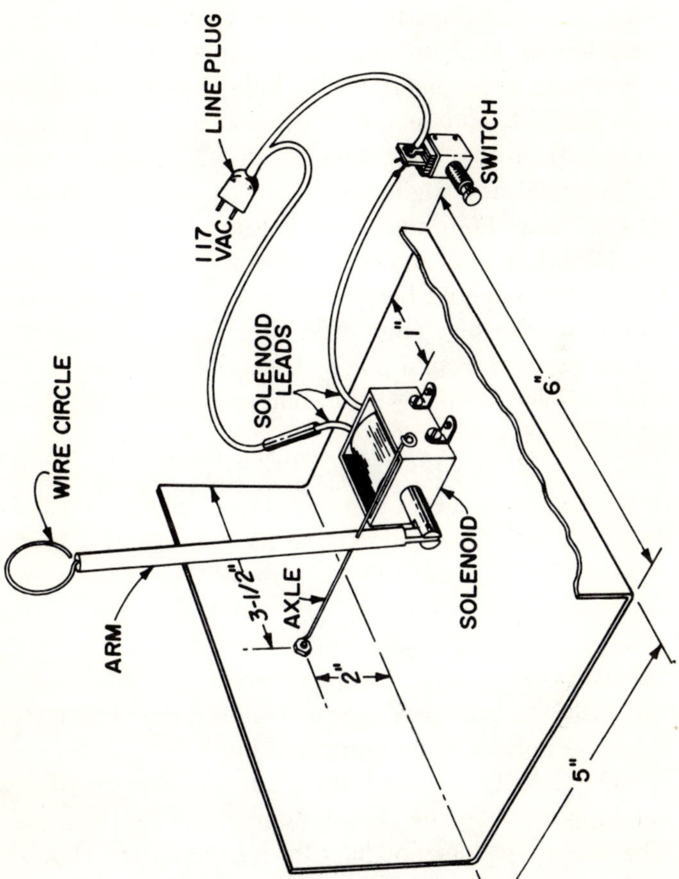

Fig. 7-3. Pictorial diagram of "The Flipper."

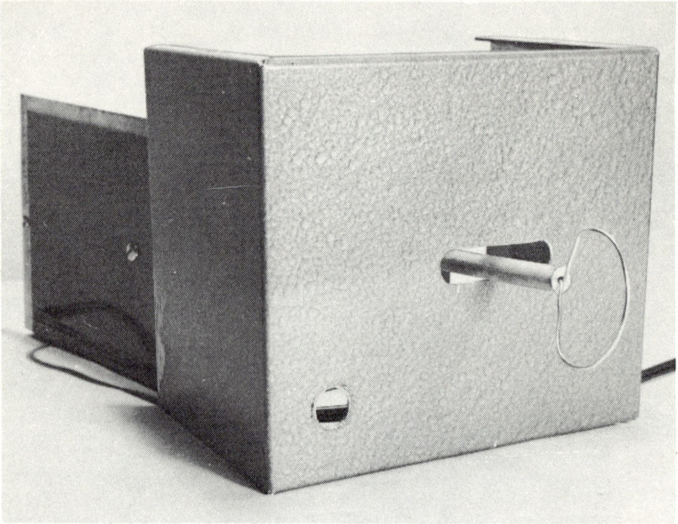

Fig. 7-4. Top view of case, showing the arm in the slot, and the hole for the push-button switch.

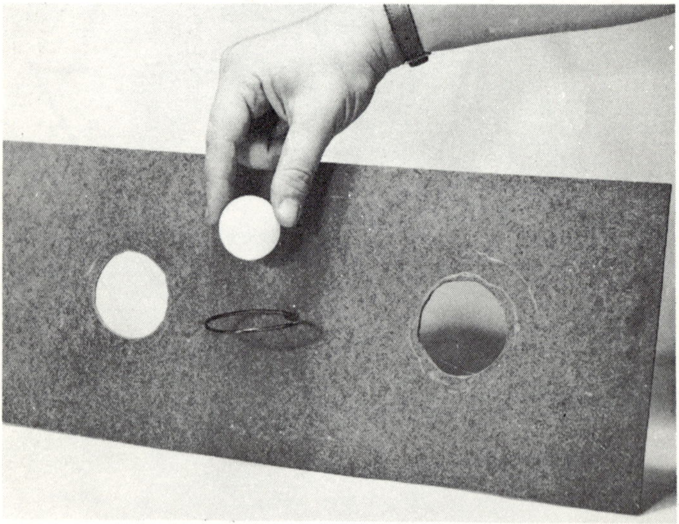

Fig. 7-5. Checking target to see if the hoop is large enough for ping-pong ball to fall through.

place. Allow long enough leads to the switch, since they must extend up to its mounting hole in the top cover of the case.

The holes in the top of the case appear in Fig. 7-4. One is for the switch, and the other is a slot through which the wooden arm protrudes. Start the slot 1¾ inches from the forward edge of the case, and run it toward the rear for another 1¾ inches. Make it just over ⅜ inch wide, to accommodate the diameter of the arm.

Bend a piece of wire into a loop big enough to support a ping-pong ball. Drill a small hole one inch deep into the top of the arm, and push the two ends of the wire loop into it. Friction will hold the loop in place if the hole is kept small.

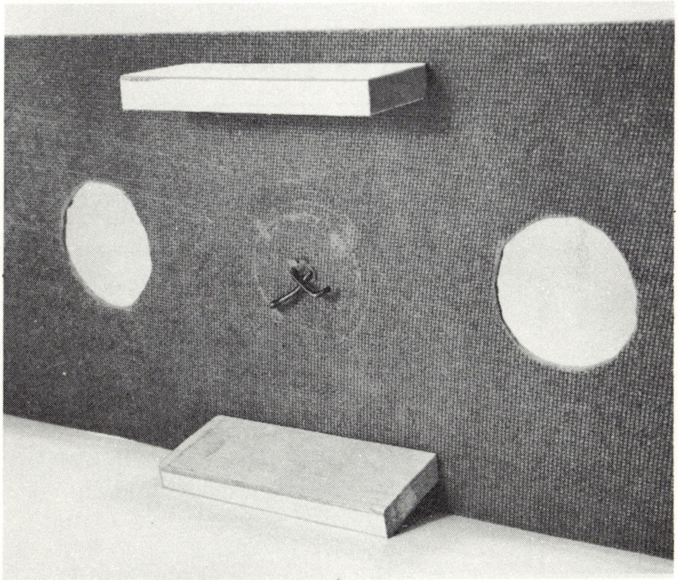

Fig. 7-6. Wood strips, mounted on the rear of the target, hold it away from wall.

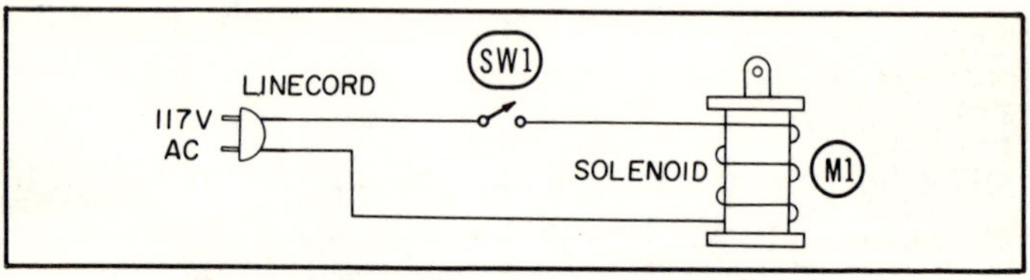

Fig. 7-7. Schematic of "The Flipper."

The target in Fig. 7-5 measures 19″ × 7″. Drill two holes 2¼ inches in diameter, and hook a wire hoop of the same dimension between them. If you intend to hang the target on a wall, keep it at least 1¾ inches away from the wall; this will leave space for the ball to drop through. One suggestion is to nail two wooden strips to the rear of the board, as shown in Fig. 7-6.

The schematic in Fig. 7-7 will serve as a guide during construction.

PARTS LIST

Item No.	*Quantity*	*Description*
SW1	1	Single-pole, single-throw switch; spring-return, push-button type.
M1	1	115-volt AC solenoid (Guardian type 60, 115VAC, intermittent; or its equivalent).
	1	AC line cord and plug.
	1	6″ × 5″ × 4″ aluminum case (*Minibox*).
	1	8¾″ length of ⅜″ wood dowel.
	2′	Stiff wire (or use coat hanger).
	4	¾″ × ⅜″ pieces of scrap metal.
	9	6-32 machine screws and nuts.
	8″	Stranded wire (switch to solenoid).
	1	19″ × 7″ piece of hardboard.
	2	⅞″ × 1¾″ × 6″ wood strips.

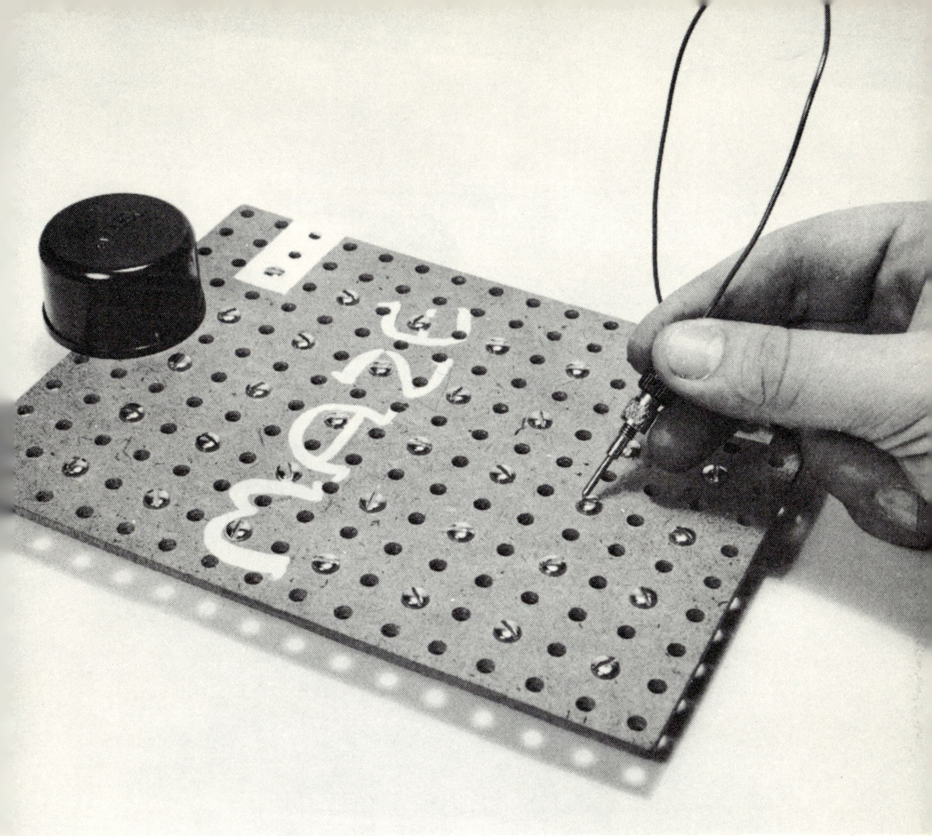

8. THE MAZE

By borrowing a tool from psychology and imparting an electronic twist to it, a game can be devised that tests the mental ability of the player. The maze has long been used by psychologists to research the learning ability of mice and thereby find the key to man's mental patterns. Through electronics, a small table-top maze can easily be constructed, replete with blind alleys and successful routes that lead the player from start to finish. A versatile feature of the game permits changing the pathway in seconds, to prevent it from growing stale after one or two plays.

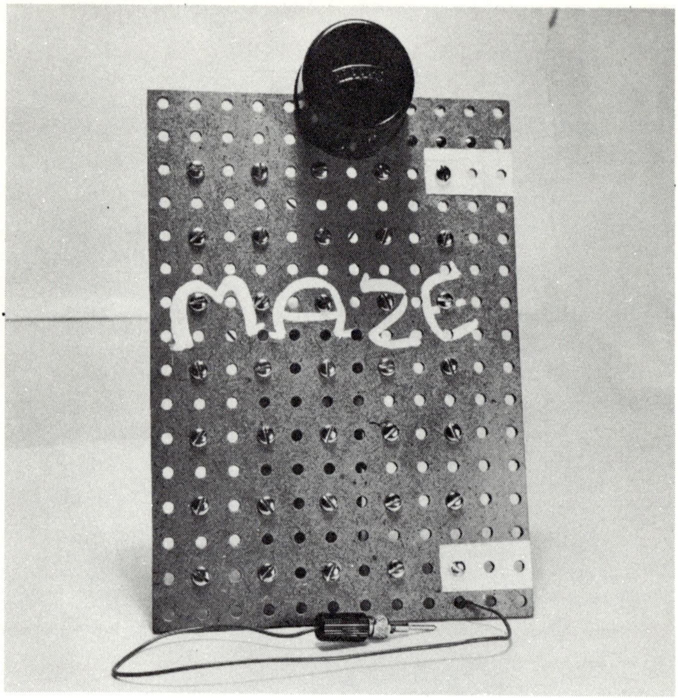

Fig. 8-1. Over-all view of topside of maze. White tab at lower right is starting point. Finish is at upper right. Buzzer is seen at top center.

The maze, as pictured above, is started by touching the stylus to the white tab at the lower right corner of the board. This is shown more clearly in the close-up view of Fig. 8-1. The player must touch a series of points on the board, in an attempt to reach the finish tab at the upper right corner. He can move in any direction—up, down, sideways, or diagonally—but only one point at a time during each move. A buzzer will sound if the correct point has been chosen. If a move is made but the buzzer does not sound, the stylus is returned to the last point that produced a tone and a new direction is at-

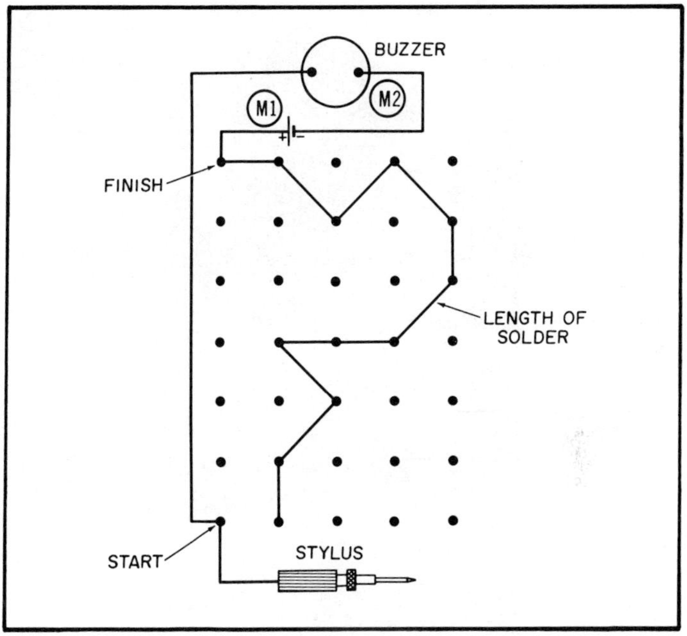

Fig. 8-2. Schematic of "The Maze."

tempted. Progress over the board is thus determined by landing on tone-producing points.

The winner is the one who can travel from start to finish in the fewest number of seconds or minutes. The game will favor the person who best profits from trial and error—in this instance, the ability to remember and thus avoid the points which yield no sound.

HOW IT WORKS

The schematic in Fig. 8-2 consists of a buzzer, battery, stylus, and contact points, all in series. Notice that several points are interconnected by a length of wire solder. If the stylus touches a point connected by the solder, the circuit from battery to buzzer closes and the

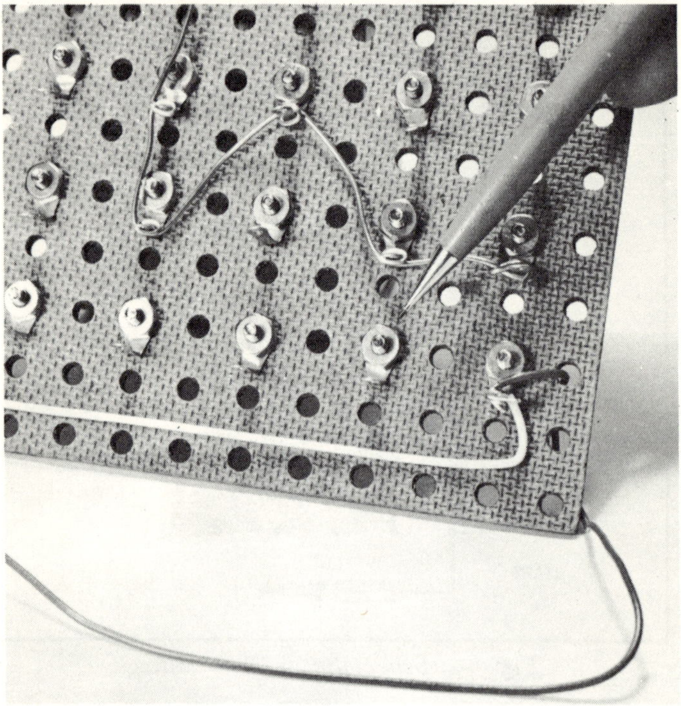

Fig. 8-3. Close-up of lower left underneath corner.

buzzer is energized. Other points produce no tone, since the circuit remains open. The correct path is not visible to the player, being out of sight beneath the board. He merely sees an arrangement of 35 screw heads, five rows horizontally and seven vertically.

CONSTRUCTION

Standard pegboard of the type having ⅛-inch holes a half inch apart is used. Cut a piece to 8″ × 6″ and insert 6-32 machine screws into the holes, as shown in Fig. 8-1. Leave one vacant hole in all directions as shown. From the underside of the board, install solder

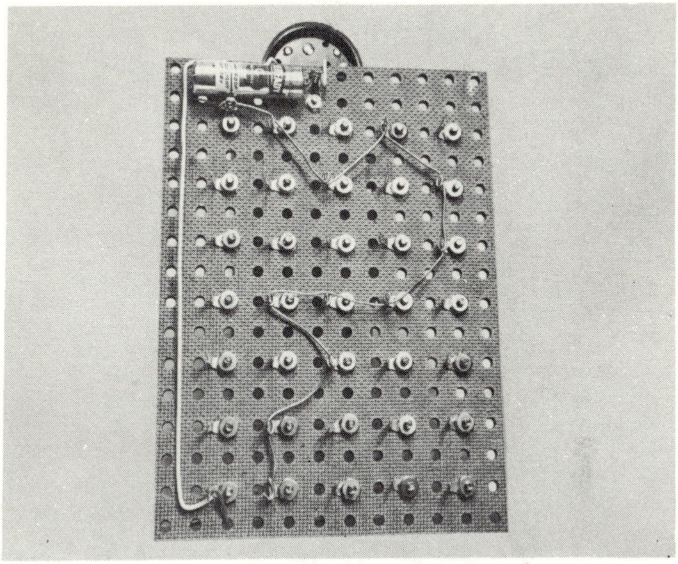

Fig. 8-4. Bottom view of completed board.

lugs over the threads of the screws and tighten in place with nuts.

The close-up view in Fig. 8-3 illustrates the underside of the board. The starting point (the terminal with the white wire connected to it) is visible at the lower right. Fig. 8-4 gives an underside view of the completed unit.

Attach the buzzer to the top center of the board with a machine screw, as shown in Fig. 8-5. Place a solder lug under the screw, and run one lead of the buzzer to this screw. Run the other lead to the solder lug on the lower left (start) terminal, as shown in Fig. 8-5.

Solder two short lengths of wire to the battery, and mount by connecting one lead to the solder lug on the buzzer-mounting screw and the other to the upper left-hand (finish) terminal. If desired, the battery can be cemented to the board. Or better yet, use a battery

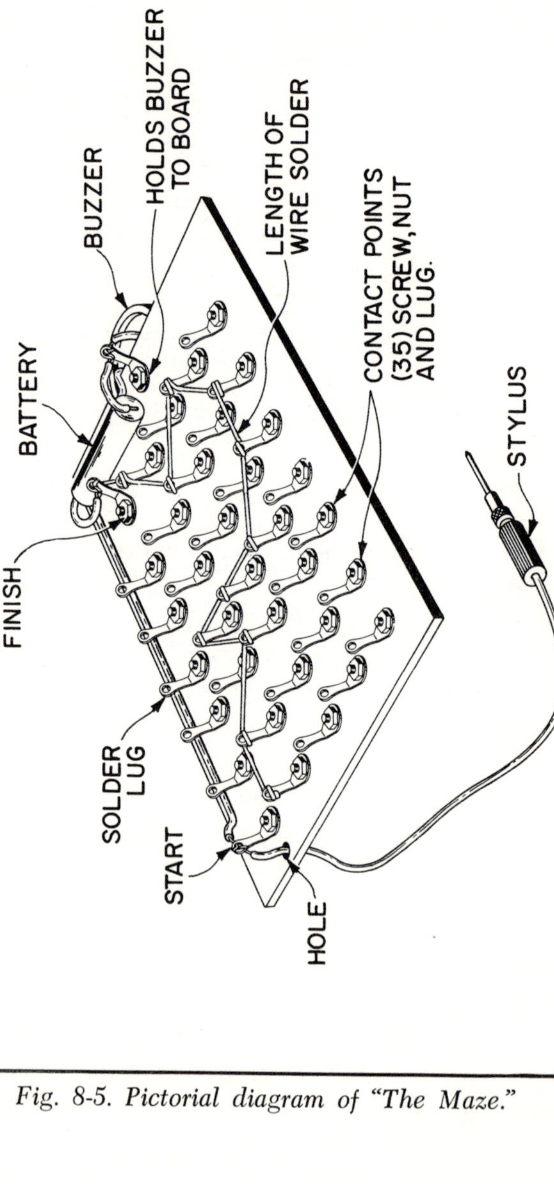

Fig. 8-5. Pictorial diagram of "The Maze."

holder and connect its terminals to the solder lugs on the buzzer-mounting screw and finish terminal.

As mentioned earlier, a strip of solder wire sets up the pathway from start to finish. No matter which group of lugs is selected, however, the wire must always start at one of two lugs—either the one just above the starting point, or the one to its right (viewed from the underside of the maze). Then loop the wire around any number of lugs, to form any pattern desired. The final tie point for the solder wire is always the top left lug, representing the finish. To assure good electrical contact, loop the wire around each lug once and pull tight. The wiring guide in Fig. 8-5 is just one of many possible pathways through the maze.

The stylus may be a piece of stiff wire fastened to a wood dowel or phone-tip jack, as shown in the model. Use a length of stranded flexible hookup wire from the stylus to the lower left-hand (viewed from rear) terminal of the board. This is the start terminal, where the lead from the buzzer was connected previously.

PARTS LIST

Item No.	*Quantity*	*Description*
M1	1	Size A flashlight cell.
M2	1	1.5-volt buzzer.
	1	8″ × 6″ piece of perforated hardboard.
	36	6-32 machine screws with nuts.
	36	Solder lugs.
	1′	Stranded hookup wire.
	18″	Solid hookup wire.
	2′	Wire solder.
	1	Phone-tip jack for stylus (or construct from wood dowel).

9. SPACE SHOT

This project features two miniature "satellites" whirling in orbit. Controlled by the players, they provide a game filled with plenty of fast action and excitement.

Both satellites are actually ping-pong balls. A stiff wire connects each ball to a wooden post rotated by a motor. As they spin, centrifugal force lifts them into an orbit at right angles to the posts, as shown above.

Assume the player who controls the dark ball is the aggressor and is intent on conquering the white ball. As both spin, there is one point where the opportunity arises. This occurs as the orbits overlap when both balls

Fig. 9-1. Top view of the board with "satellites" at rest in front of the posts.

pass over the top center of the board and are traveling fast enough to lie straight out from their posts. During this instant, physical contact is possible, and the player controlling the dark ball can score a strike. The player with the white ball tries to evade his opponent and thus prevent him from scoring.

The game hinges on the players' being able to vary the height of each ball's orbit. As long as the player button is held down, the ball simply spins at its maximum height. By tapping the button (like a telegraph key sending a code), it is possible to vary the height. This is the skill factor. One rule should be imposed—both players must keep their "satellites" in motion at all times. Fig. 9-1 shows the two balls at rest.

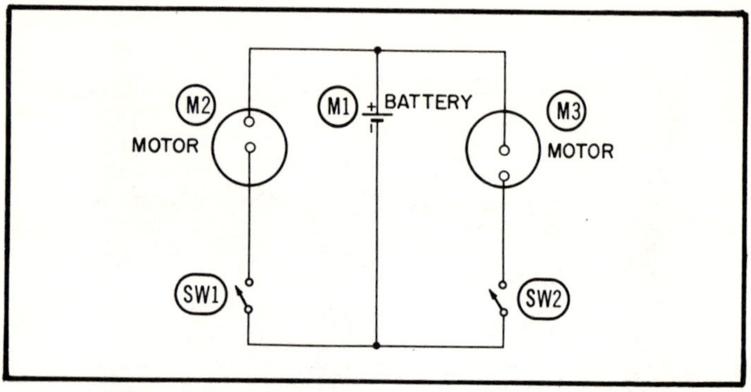

Fig. 9-2. Schematic of the "Space Shot."

HOW IT WORKS

A schematic of the "Space Shot" appears in Fig. 9-2. Two small DC motors rotate the posts whenever the buttons are depressed. A single battery provides the power. The direction in which the motor rotates depends on how its terminals are wired to the battery. A more desirable effect is achieved when the balls spin in opposite directions. In the model, the dark ball spins in a clockwise direction and the white one spins counterclockwise.

CONSTRUCTION

Fig. 9-3 shows the underside of the board. The width and length measure 12″ × 8″. Two wood strips serve both as mounting legs and as supports for the motors. Fig. 9-4 shows how the motors are mounted on the board. Drill two ½-inch holes for the shafts, and attach each motor to the sides of the wood strips with two wood screws.

On the topside, the two posts are now fitted to the motor shafts. Cut two pieces of ⅜-inch wood dowel four

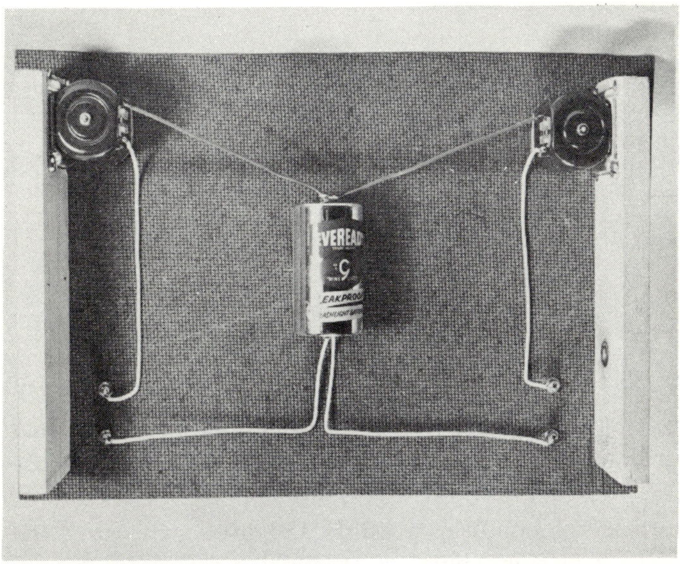

Fig. 9-3. Bottom view of board.

Fig. 9-4. Close-up of motor mounting.

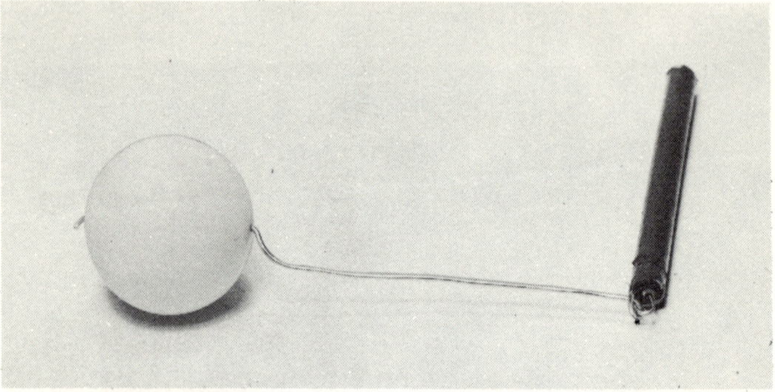

Fig. 9-5. View of wood dowel with ball attached.

inches in length. Drill a small hole about half an inch deep into the ends of each dowel. Since the motor shafts are pressed into these holes, it is important that they be as near the center as possible. Otherwise, the posts will wobble as they spin.

Drill similar holes in the other ends of the dowels. Press a U-shaped wire (made from a paper clip) into each hole. These will form eyes to receive the wire lengths from the balls.

Now prepare each ball and its wire as follows. With a thumbtack, puncture two holes along the "equator" on opposite sides of a ping-pong ball. Pass a 5½-inch wire through the holes, and bend up the end about one-fourth inch. This is done to prevent centrifugal force from sliding the ball off the wire. Curve the wire slightly, next to the hole that will face the post, to keep the ball from sliding in this direction. Form the free end of the wire into a small loop and insert into the eye in the wood dowel. The completed ball and dowel is shown in Fig. 9-5. Make certain the ball is free to rise and fall as it spins around the post. Bend the loop if any binding is noticed.

Repeat these steps for the other ball.

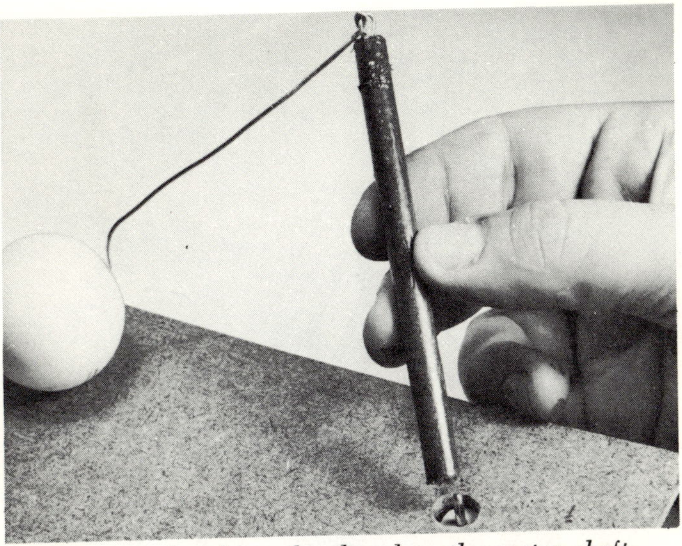

Fig. 9-6. Attaching the dowel to the motor shaft.

Installation of a wooden post is shown in Fig. 9-6; the lower end of the dowel is pressed onto the motor shaft. It is a good idea to place one hand under the motor, to support it as you are fitting the dowel in place.

Fig. 9-7. Detail of switch-button construction.

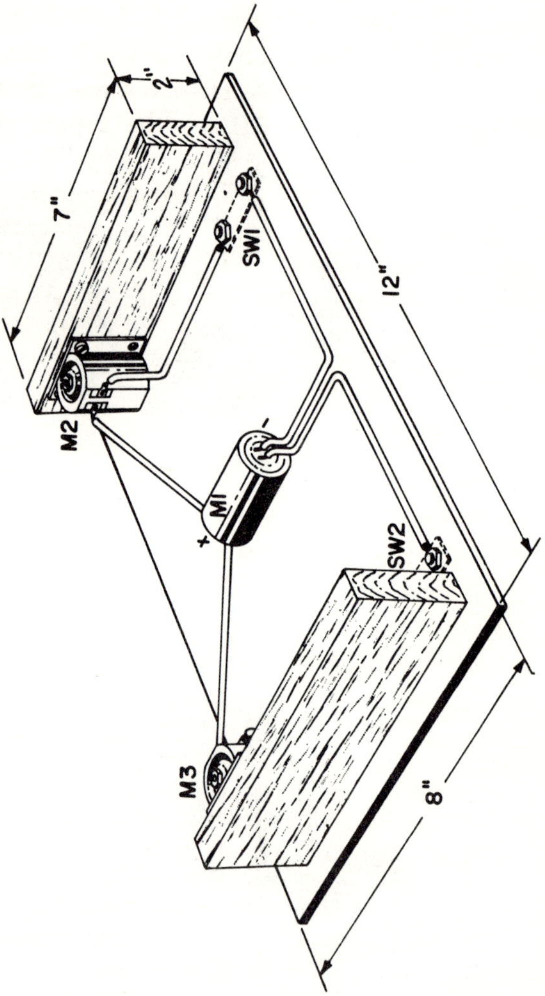

Fig. 9-8. Pictorial diagram of the "Space Shot."

Fig. 9-7 shows a close-up view of the switch construction. Cut two strips from a piece of copper (sold as copper flashing in hardware stores) and drill a hole at one end of each piece. Drill two holes on each side, near the front of the board. Insert 6-32 machine screws through the front holes and tighten. Attach the copper strips with two more 6-32 screws, as shown in Fig. 9-7.

The battery is mounted underneath the board, near the center, as shown in Fig. 9-8. It can be cemented to the board, and the hookup wires soldered to the terminals as shown in the model; or you can use a battery holder and attach the hookup wire to its terminals. Connect the other ends of the battery wires to the motors and switches. Then add hookup wire from the switch terminals to the motors, as shown in Fig. 9-8. The wires can be placed under the nuts on the switch terminals and the nuts tightened to hold in place. If the screws are plated, you can wrap the end of the wire around the screw and touch with solder.

PARTS LIST

Item No.	*Quantity*	*Description*
SW1, SW2	2	Switches (see text).
M1	1	Size D flashlight cell.
M2, M3	2	1.5-3–volt miniature DC motors.
	2	Ping-pong balls.
	2	4½″ × ⅜″ dowels.
	1	8″ × 12″ piece of hardboard.
	2	⅞″ × 2″ × 7″ wood strips.
	2	5½″ lengths of stiff wire (or use a coat hanger).
	4	Wood screws.
	4	6-32 machine screws with nuts
		Hookup wire.

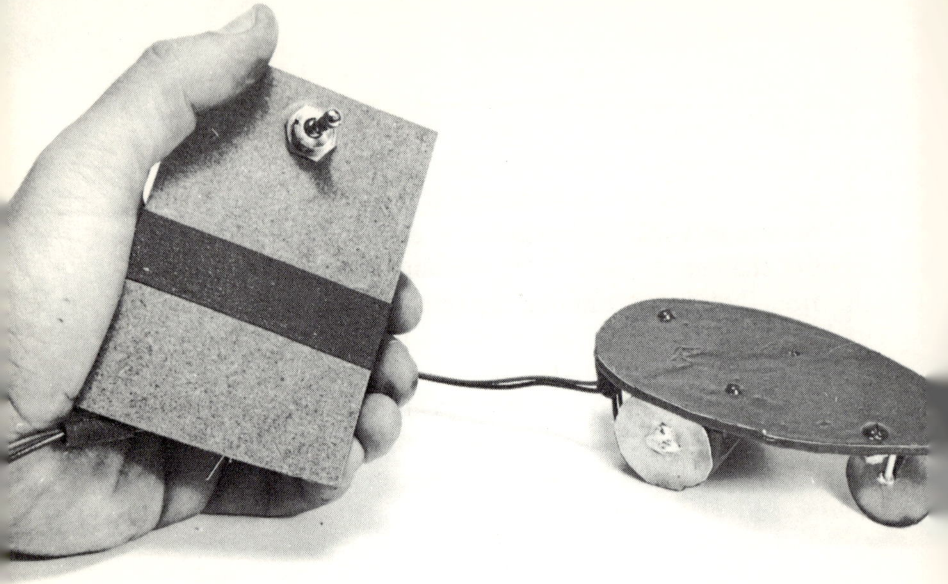

10. MOVING MOUSE

Intended for the amusement of a young child, the "mouse" is a toy that grown-ups will be unable to resist. As it scoots across a living-room floor, it can be stopped or backed up by remote control. An optional version, to be described later, can be turned to dodge obstacles. The two basic sections are shown in the above photo. The hand is holding the control unit and the mouse is at the end of a control cable at the right. The switch on the control provides the forward-reverse function. When power is applied, the mouse proceeds across the floor. The reversing switch is flipped whenever the mouse reaches the end of the cable or its path is blocked.

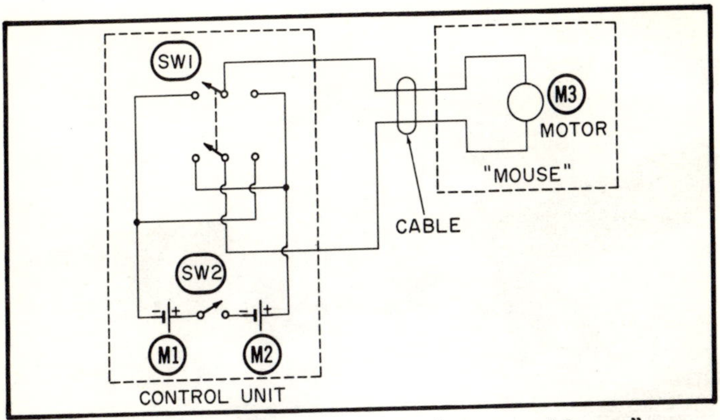

Fig. 10-1. Schematic of the "Moving Mouse."

HOW IT WORKS

Two batteries in the control unit send current, through the cable, to a small DC motor underneath the mouse. (See schematic, Fig. 10-1.) The switch reverses the polarity of the voltage applied to the DC motor and thus its direction of rotation.

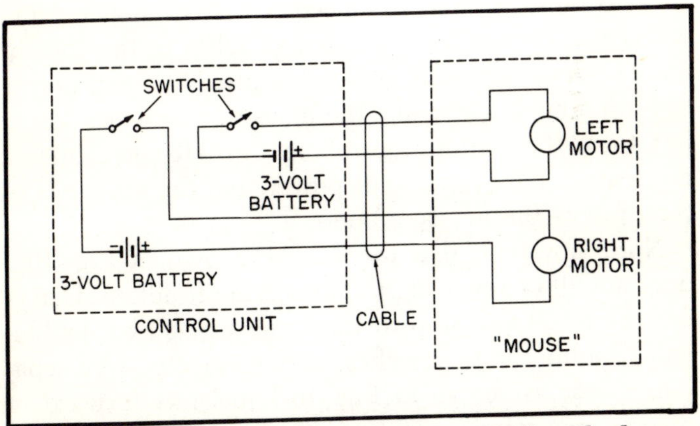

Fig. 10-2. Schematic of the "Moving Mouse" with the optional two-motor drive for steering.

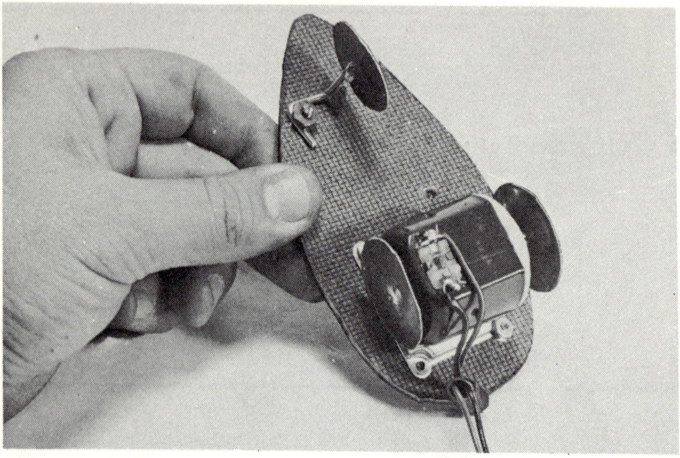

Fig. 10-3. Underside view of mouse. The drive motor near the lower edge rotates the rear wheels and the front wheel, near top, spins freely.

It is possible to make the mouse more maneuverable by using two motors, one for each wheel. In this way, you can steer the mouse to the right or left. This feature is shown in the schematic in Fig. 10-2, for those who wish to build a more advanced toy. Note that a four-wire cable is required, plus two switches in the control unit. Both switches are pressed to provide equal power to both wheels and make the mouse go forward. For a left turn, the button controlling the left-wheel motor is released. This shuts off its power, and the other wheel then pivots the mouse around.

Note, however, that the version shown in Fig. 10-2 does not allow reversing the direction of motion. Carrying the revision a step forward and using two double-pole, double-throw switches, each connected to a separate motor as shown in Fig. 10-1, permits forward or reverse movement and turns while traveling in each direction.

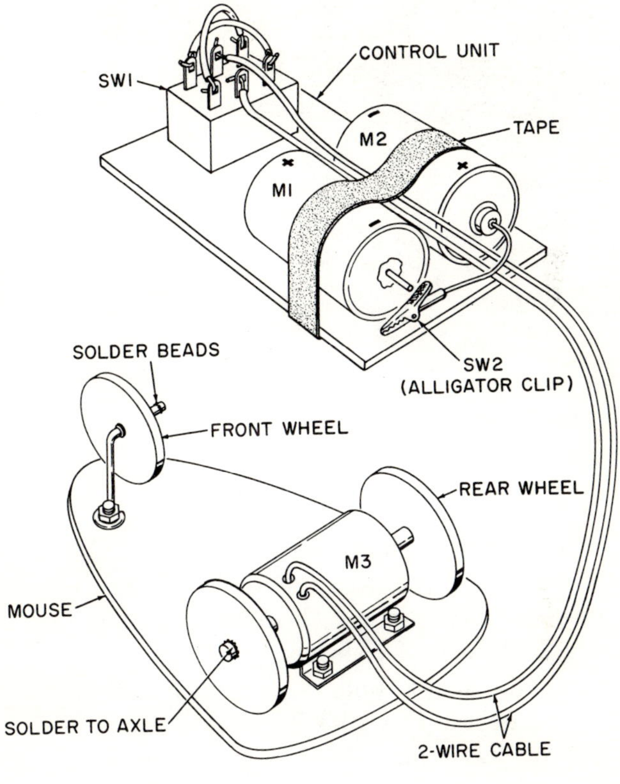

Fig. 10-4. Pictorial diagram of the "Moving Mouse."

Fig. 10-5. Underside of control unit.

CONSTRUCTION

Cut two pieces of hardboard, one to 4″ × 2½″ for the control unit and the other about five inches long and in the shape of a mouse. This dimension will vary according to the size of your motor.

Secure the motor under the mouse with nuts and bolts, as shown in Fig. 10-3. The wheel diameter should be slightly *less* than twice the clearance between the motor shaft and the underside of the board. Cut the three wheels from copper, drill a center hole in each, and solder two of them to the ends of the motor shaft. Bolt a piece of stiff wire to the board and bend it down for the axle. To keep the wheel in place, drop two beads of solder on the axle (on either side of the wheel.) A pictorial diagram of both units is given in Fig. 10-4.

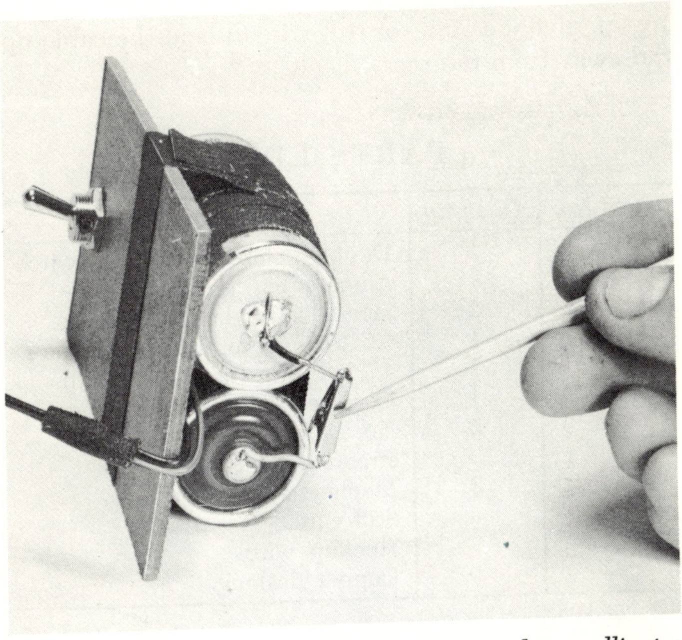

Fig. 10-6. Detail of on-off switch. Pointer shows alligator clip for connecting battery terminals.

Fig. 10-5 shows the underside of the control unit. Drill a hole and mount the reversing switch. Connect jumper wires across the switch terminals as shown in Figs. 10-4 and 10-5. Tape the batteries to the board and connect the leads from the battery to reversing switch terminals. Solder two short leads to the remaining battery terminals and solder an alligator clip to one lead as shown in Fig. 10-6. This clip becomes the on-off switch (SW2). To turn it off, just disconnect the clip.

The cable from the control unit to the mouse may be any convenient length (around five feet). Flexible, stranded wires will least interfere with the movement of the mouse. No. 20, or the thinner No. 22, is recommended. If the mouse trips over the cable while backing up, give it a "tail" of stiff wire to hold the cable up and away from the rear wheels.

PARTS LIST

Item No.	*Quantity*	*Description*
SW1	1	Double-pole, double throw switch.
SW2	1	Alligator clip (see text).
M1, M2	2	Size D flashlight cells.
M3	1	1.5- to 6-volt miniature DC motor.
	1	4½" × 2½" piece hardboard.
	1	5" piece hardboard (for "mouse")
	5′	No. 20 or No. 22 two-wire stranded cable.
	3	Machine screws and nuts.
		Stiff wire.
		Hookup wire.
		Copper flashing.

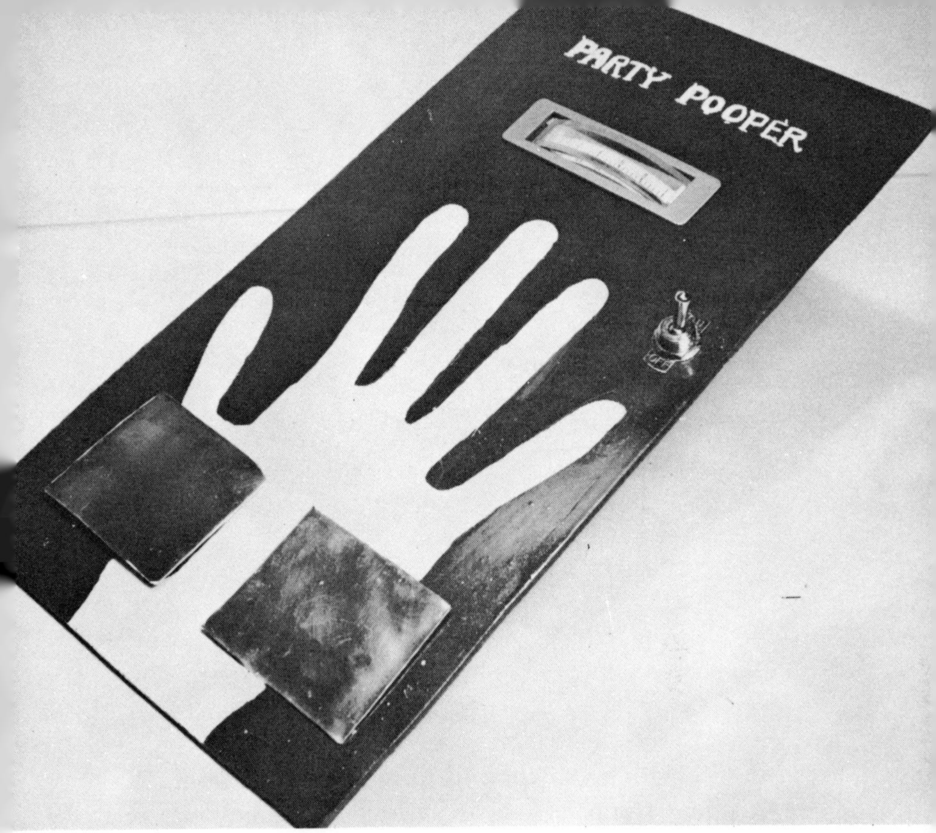

11. PARTY "POOPER"

Just when the party is about to "poop" prematurely, it's time to make an announcement that might sound something like this:

"Ladies and gentlemen! Through the miracle of electronics, we are now going to attempt an experiment in 'electroanalysis' "!

After several appropriate remarks in this vein, the host brings out the "Party Pooper" shown above, a weird

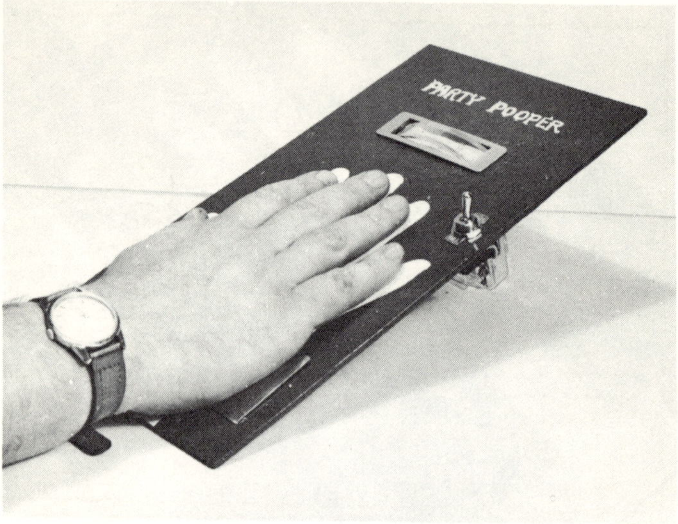

Fig. 11-1 Game being used as a lie detector.

device with a ghostlike white hand painted on it. What happens from here on is unpredictable, but is sure to enliven the dullest gathering.

The "Pooper" entertains in a variety of ways. It will act as a lie detector or "kiss-o-meter," and will even test the "sincerity" of a handshake! While the results cannot be certified 100%, they will provoke loads of laughter.

The first method of using the unit is shown in Fig. 11-1. A person is invited to place his hand over the painted hand on the board. At first the needle on the meter will climb, and then come to rest somewhere on the scale. The host now asks the person a series of questions, pausing for several seconds between them. If the needle moves to a higher reading, the host humorously announces that a falsehood has been uttered. This is the lie-detector function of the "Pooper." Although not as precise as the Keeler polygraph, it is fine for parlor use.

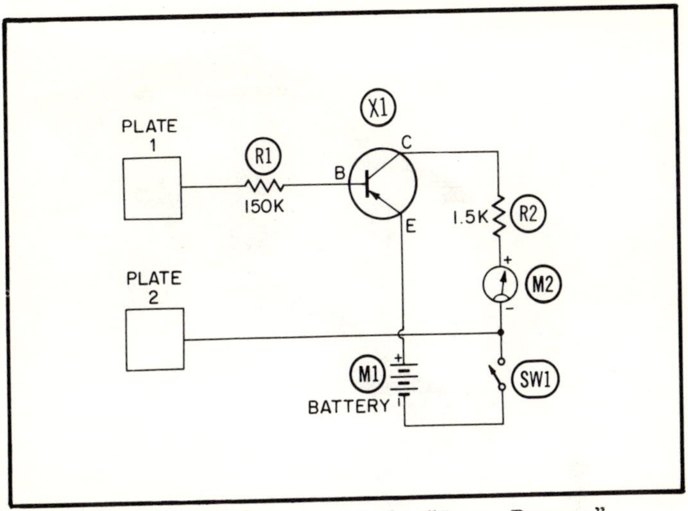

Fig. 11-2. Schematic of the "Party Pooper."

As a "kiss-o-meter" or handshake tester, the unit is used in a slightly different fashion. Two "volunteers" are selected. Each is instructed to place his or her left index finger on one of the plates in the area of the "palm" on the board. As they kiss or shake hands, the meter is consulted. The higher the reading, the more engenuine the sincerity of whatever activity they are engaged in!

HOW IT WORKS

The device is based on the skin resistance of the human body. Measured in ohms, it is rather high. But a small amount of moisture is always present on the surface of the skin, and varies with a person's state of anxiety. This moisture sets up a conductive pathway that allows sufficient current to flow and energize the circuit. There is no shock hazard, since the current is at the microampere level and the battery is a safe 9 volts.

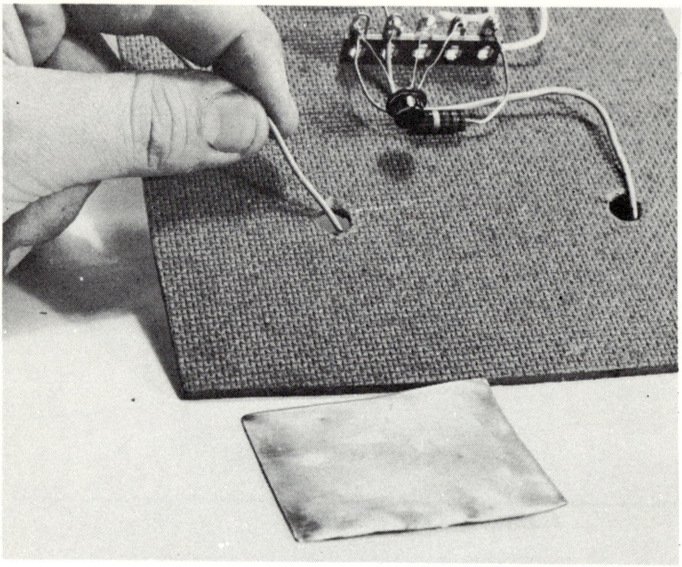

Fig. 11-3. Installing copper plates to the board. Wire is inserted into hole and soldered to plate.

The schematic in Fig. 11-2 explains the theory behind this device. The two plates (copper squares in the palm section of the board) are connected to the transistor base and negative battery terminal, respectively. When a hand is placed across the plates, its conductivity allows current from the battery to reach the base, or control element, of the transistor. The transistor amplifies the microampere current flow across the person's palm until it is high enough to produce a reading.

The variable element is the resistance of the hand. The more moisture on the skin, the lower the resistance between plates. The battery voltage drives more current to the base of the transistor, and as a result, the meter reading increases. The same situation exists when two persons touch the plates with their fingertips and then kiss or shake hands. The resistance path is now

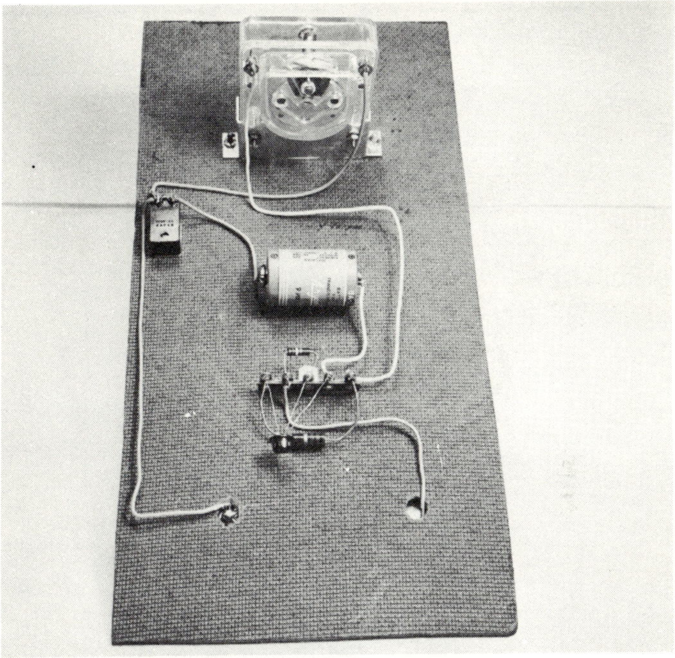

Fig. 11-4. Over-all view of board's underside.

diverted through their bodies. The transistor is sensitive enough to make the circuit work in spite of the fairly long path between plates.

In the handshake method, the firmer the grip, the higher the meter reading. This is due to the lower total skin resistance as the pressure of one palm against another increases the contact area between the two persons.

CONSTRUCTION

Start by cutting a piece of board to 7″ × 12″. Paint it a dark color. After it has dried, place your right hand on the board, in the position suggested by the photos.

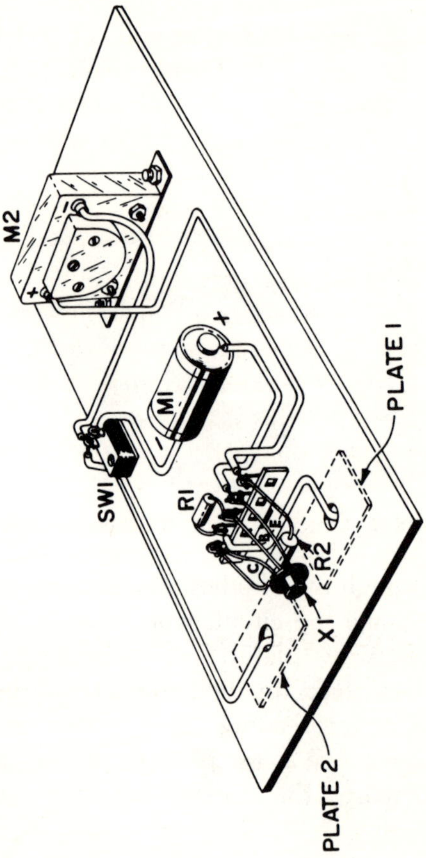

Fig. 11-5. Pictorial diagram of the "Party Pooper."

Then trace its outline in pencil, and brush it in with white paint.

The two plates mounted in the palm area of the board are next. Cut two 2-inch squares from a sheet of copper, and position them on top of the board. Mark off two holes for the wire leads that will connect to the plates from the bottom of the board. (Note Fig. 11-3.) Push each wire through its hole and solder it to the underside of the copper plate. After this is done, cement the plates in place.

An over-all view of the board's underside appears in Fig. 11-4. An "edgewise" type of meter was used here, but is not absolutely necessary. Any round or square meter capable of reading 0-1 milliampere DC full scale will do. Avoid too small a meter, however; its poor visibility will reduce the effectiveness of the game. Cut a hole in the board the size of the meter dial. Drill two holes and attach the meter to the board with machine screws.

The wiring guide in Fig. 11-5 gives the connections for the other components. Mount a strip underneath the board as shown. Then connect the two resistors and

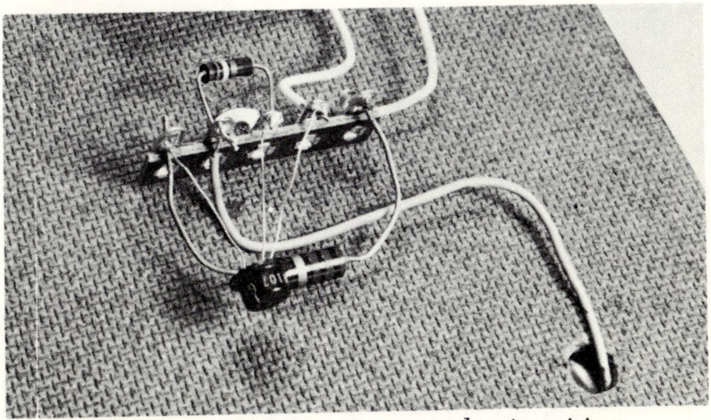

Fig. 11-6. Close-up of terminal-strip wiring.

the transistor, battery, and switch. Fig. 11-6 is a close-up view of the wired terminal strip. The battery may be glued to the board as shown in the model, or a battery holder may be used.

PARTS LIST

Item No.	*Quantity*	*Description*
X1	1	Transistor, pnp, germanium, audio (2N107, TR-05, HEP253, ECG102, SK3003, GE-2, AT10N).
R1	1	150K, ½-watt resistor.
R2	1	1.5K, ½-watt resistor.
SW1	1	Single-pole, single-throw toggle switch.
M1	1	9-volt battery.
M2	1	0-1 DC milliammeter.
	1	7″ × 12″ piece of hardboard.
	2	2″ × 2″ copper plates.
	1	5-lug terminal strip.
	3	6-32 machine screws and nuts.
		Hookup wire.

12. COUNTDOWN

A keen sense of timing singles out the winner of "Countdown," a game built around a one-transistor circuit. It can be played by two persons, or an individual can pit his skill against the unit's electronic abilities. The idea is to see who can guess the passage of time more accurately—you or your opponent.

The playing arrangement is shown in Fig. 12-1. The two players hold their fingers on the push buttons marked "A" and "B" (just below the two pilot lamps, near the sides of the board). Depressing either button will light its bulb at any time.

Fig. 12-1. Front panel view of the completed unit.

The game is begun by setting a time period with the knob in the center. The pointer may be rotated opposite any number from *2* to *20*. These calibrations represent time in seconds. The circuit is then "armed" by flipping the "Ready" switch down for several seconds, and then up again. The instant this is done, the players start counting mentally. During the countdown, the bulb at the top of the board stays lit; then it automatically goes out at the end of the period shown by the knob pointer. The winner of the round is the player who presses his button (and thus lights his lamp) last *before* the timing bulb goes dark. If the period is over and neither player has pressed his signal, the game is considered a draw. The next round commences by setting the pointer to another number and holding the "Ready" switch down momentarily. Timing begins the instant the "Ready" switch is flipped up. When the

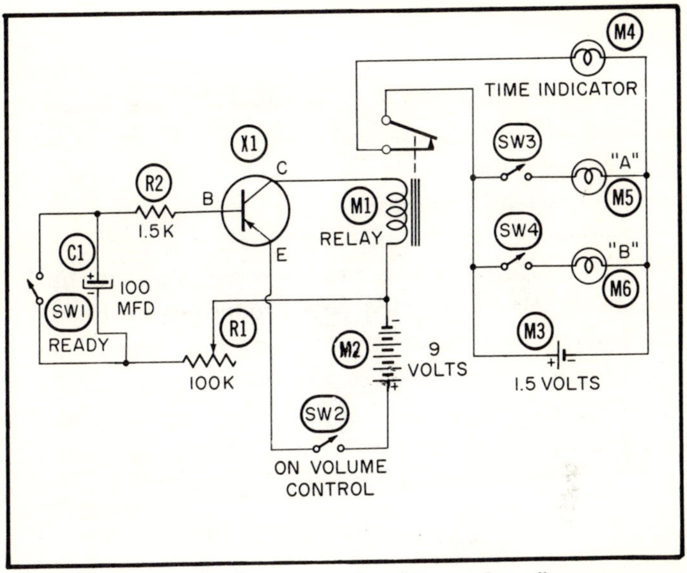

Fig. 12-2. Schematic of the "Countdown" game.

game is not in use, the knob is rotated fully counterclockwise. A built-in switch then disables the power.

HOW IT WORKS

The electronic action of the circuit can be traced from the schematic in Fig. 12-2. The negative battery terminal supplies a source of electrons. These flow through potentiometer R1 and onto the upper plate of C1, a large capacitor. As C1 charges up, its lower plate becomes increasingly positive. This voltage is applied to the base element of the transistor, reducing the current flow from collector to emitter. A relay in the collector circuit remains energized (on) until the base becomes positive enough to cut off most of the collector current. At this point, the relay contacts open and the time indicator lamp (M4) goes dark.

The variable time feature of the circuit is the potentiometer. When its resistance is changed by rotating the knob on the panel, the capacitor charging time also is changed. The players' bulbs "A" and "B" are on separate circuits and go on and off as switches SW3 and SW4 are opened or closed.

Another aspect of the circuit is the "Ready" switch, SW1. Note that it is across the two plates of the capacitor. In the closed position (corresponding to "Ready"), SW1 shorts out the capacitor and thus discharges it completely. Flipped up to the open position, it frees the capacitor to accept another surge of electrons for the next round of the game.

CONSTRUCTION

A board measuring 7½″ × 5″ serves as the chassis. Fig. 12-3 shows the mounting of the various components. First drill three holes for the bulbs, fit the rubber grommets in the holes and slip the bulbs into the grommets (friction will hold them in place). Next drill mounting holes for switches SW1, SW3, and SW4, and potentiometer R1. (The size for the holes will vary, depending on the construction of the components selected.) Mount SW1, SW3, SW4, and R1 to the board. Drill mounting holes for the relay M1 and the 4-lug terminal strip and attach with machine screws.

Now that all major parts are mounted, proceed with the wiring connections and installation of the remaining components. Use the pictorial diagram in Fig. 12-3 and the view of the underside of the completed chassis in Fig. 12-4 as guides. Make sure you wire the batteries and the capacitor into the circuit in the proper direction.

Attach the two wood strips to the edge of the chassis as shown in Fig. 12-4. It is best to taper one side of the

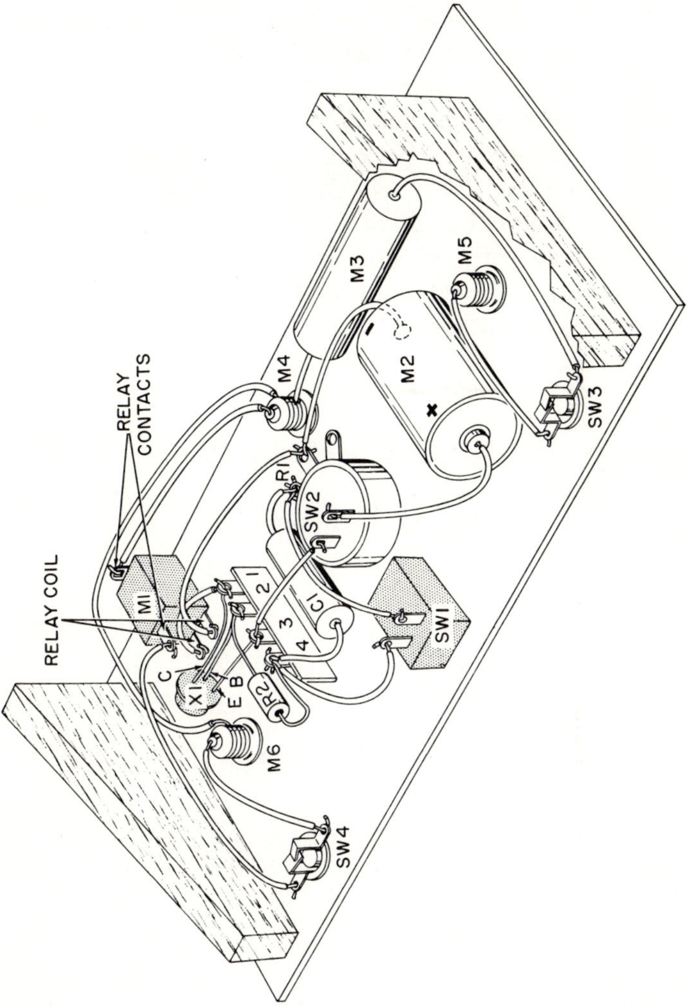

Fig. 12-3. Pictorial diagram of the "Countdown" game.

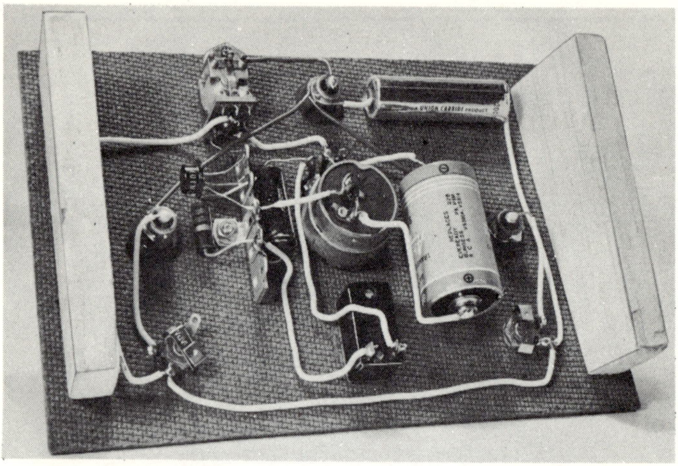

Fig. 12-4. Underside view of the completed game.

strips so the completed unit will sit at an angle for playing.

After the game is wired, some calibration (marking the timing numbers on the front panel) is in order. Your completed model will *not* necessarily agree with the numbering shown on the unit—there is an unavoidable variation in parts which prevents this. If no wiring error exists, however, any model should fall into the approximate timing range shown.

The best place to start the calibration is at the relay. First, turn the knob so it points toward the top edge of the panel. Flip the "Ready" switch down and then up, to set the circuit in operation. If the top timing bulb does not light, readjust the relay spring slightly. Notice in Fig. 12-5 that it is anchored by a small metal tab. With the pointer knob straight up and the "Ready" switch also up, gently bend the tab with the tip of a screwdriver until the timing bulb lights. This relieves some of the spring tension and allows the relay to close properly.

Fig. 12-5. Adjustment of the relay spring.

The final step is painting the timing indications on top of the panel. Turn the knob to various positions and, using a watch with a second hand, determine the number of seconds the bulb takes to go out. Then paint these markings at the appropriate points.

PARTS LIST

Item No.	*Quantity*	*Description*
X1	1	Transistor, pnp, germanium, audio (2N107, TR-05, HEP253, ECG102, SK3003, GE-2, AT10N).
C1	1	100-mfd, 15-volt, electrolytic capacitor.
R1	1	100K potentiometer, with switch.
R2	1	1.5K, ½-watt, resistor.
SW1	1	Single-pole, single-throw toggle switch.
SW2	1	Single-pole, single-throw switch (on potentiometer).
SW3, SW4	2	Single-pole, single-throw (spring return) push-button switches.

(continued)

PARTS LIST (cont'd)

Item No.	*Quantity*	*Description*
M1	1	5000-ohm, 1.8 ma relay (Lafayette *Little Jewel*, or equivalent).
M2	1	9-volt battery.
M3	1	Size A flashlight cell.
M4, M5, M6	3	No. 49 pilot lamps.
	1	7½" × 5" piece hardboard.
	1	4-lug terminal strip.
	3	⅝" rubber grommets.
	2	⅞" × 3" × 5" wood strips.
		Machine screws.
		Hookup wire.

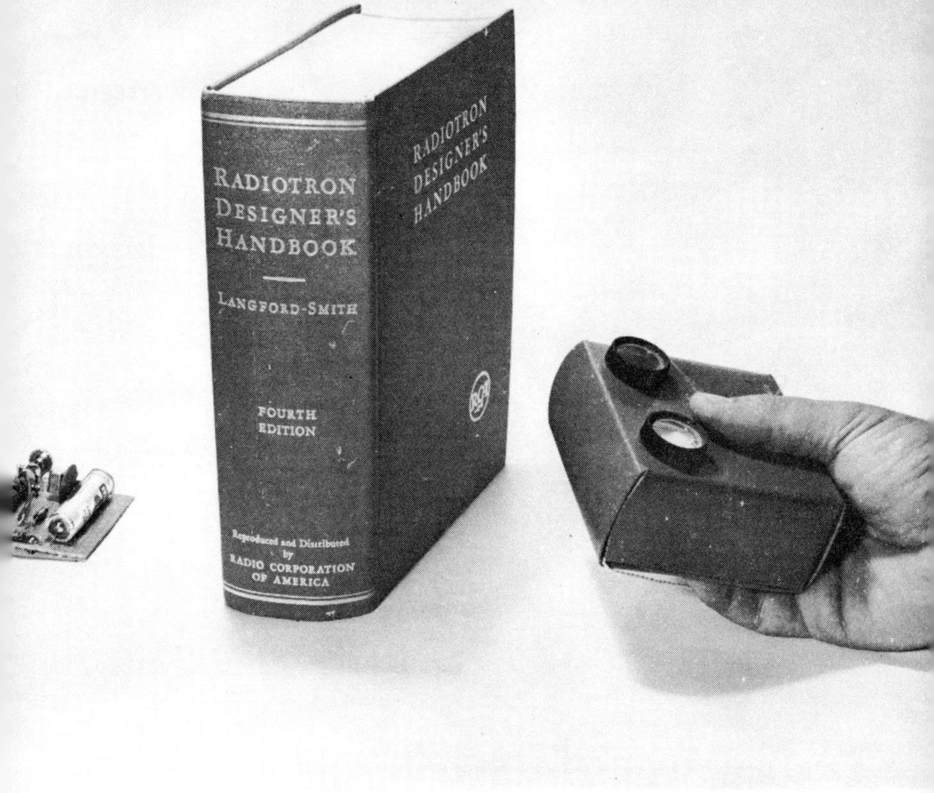

13. FIND THE AIRPORT

Principles used in navigating real aircraft are incorporated into the game of "Find the Airport." They can be re-created in your living room in a project that combines interest with instruction. During the play, an "airplane" is flown under conditions of zero visibility, in an attempt to find the home airport. The successful flight is the one that reaches the airport before the fuel tanks run dry!

An over-all view depicting how the game is played is given above. The "pilot" manipulates a transistor porta-

Fig. 13-1. Top view of the completed oscillator.

ble radio representing his radio direction-finding equipment (RDF). The beam he is flying is radiated by the airport (the small unit concealed behind the book). By employing the proper RDF techniques, the pilot will be able to follow the radio beam and land safely.

The photograph gives the game's basic idea only. Normally the "airplane" starts its flight at a much greater distance, and the airport is more effectively concealed (to simulate "zero" visibility). Of course, the airport (radio beacon) is hidden while the pilot is out of the room.

HOW IT WORKS

The "airport" is a tiny transistor oscillator which radiates a signal throughout the room. Power is just a few milliwatts, but is sufficient for a range of about fifteen to twenty feet. The oscillator frequency is adjusted to the low end of the standard broadcast band. Fig. 13-1 gives a view of the completed unit.

The "airplane" is an ordinary transistor portable radio. It must have at least five transistors. Crystal receivers, or sets with less than five transistors, are not sensitive enough to pick up a usable signal beyond one or two feet. The game is most effective when the pilot can fly his craft anywhere in the room and still hear the beacon signal.

A peculiarity of the transistor radio allows it to sense the direction of the beacon. The loopstick in these sets is directional. Thus, if you tune in a broadcast station and rotate the radio in a complete circle, you will find that the program drops out at two distinct points. Called "nulls," these points occur whenever the antenna is at right angles to the transmitting station, because the loop is least sensitive in this direction. There are, of course, two nulls. In actual RDF equipment, one null is eliminated by an additional element called a "sense" antenna. Only the remaining null is used for determining direction. In this game we have no sensing antenna, but the next section gives a technique for finding the correct null anyway.

CONSTRUCTION

The oscillator is built onto a piece of perforated board cut 2½ inches square. After determining the location of major parts, using Figs. 13-1 and 13-2 as guides, thread the wiring through the board holes to keep everything securely in place. Fig. 13-3 reveals how the circuit should look on the underside. The battery is held in place by two leads soldered to its terminals.

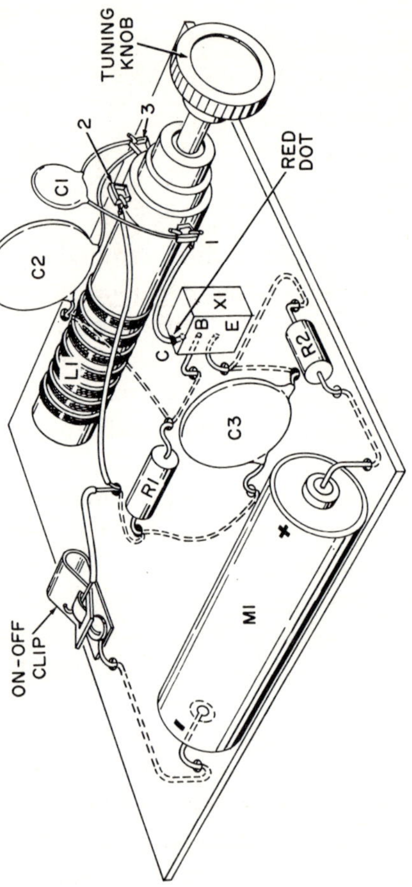

Fig. 13-2. Pictorial diagram of the oscillator.

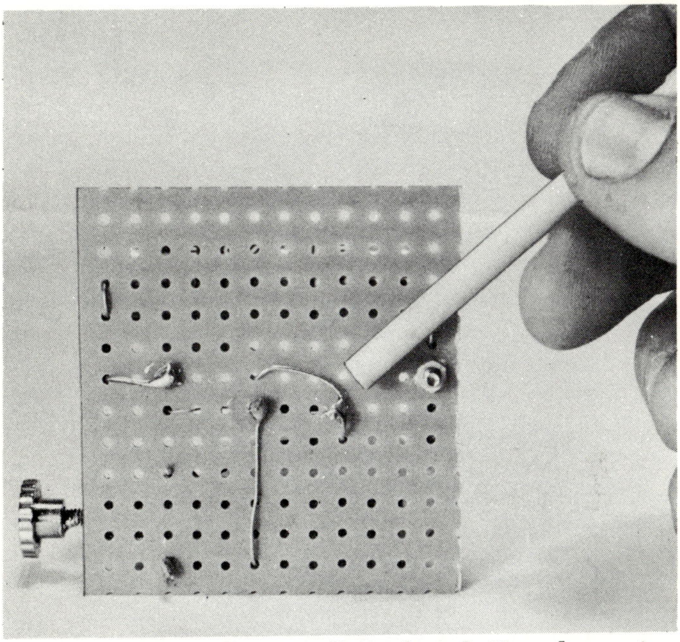

Fig. 13-3. Underside of oscillator board. Note how wires are threaded through holes to secure the parts.

The on-off switch is simply a Fahnestock clip bolted to the board, with the appropriate wire soldered to it. (See Fig. 13-4.) To turn the oscillator on, merely insert the free switch wire into the clip.

The schematic in Fig. 13-5 gives the circuit. Be sure to use a 3-tap transistor antenna coil. The numbering may not be the same as shown here, but it can be checked out with an ohmmeter. The lowest resistance should occur between the coil terminals connected to capacitors C3 and C2. The terminal with the greatest resistance to the remaining two terminals should be connected to the collector (C) of the transistor. Incidentally, any coil used must have an adjustable core for tuning purposes.

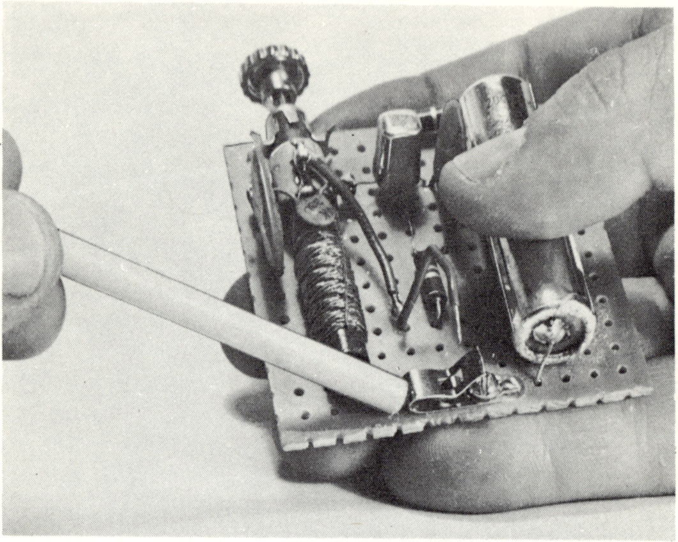

Fig. 13-4. Fahnestock clip used as on-off switch.

PRELIMINARY SETUP AND OPERATION

After constructing the oscillator, tune the transistor radio to the lowest frequency on the dial (below 550 kc). The dial should rest against its tuning stop. (This prevents broadcast stations from interfering with the game.)

Hold the oscillator about four or five feet from the radio and start tuning coil L1. When the oscillator and radio frequencies are the same, you will hear a strong rushing sound in the speaker. Avoid tuning the oscillator to the intermediate frequency of the radio. This is checked by tuning the receiver across the broadcast band. If a whistling sound is heard in the background on all stations, try another coil setting.

To determine the directional characteristics of your radio, set the oscillator in operation and place it atop a

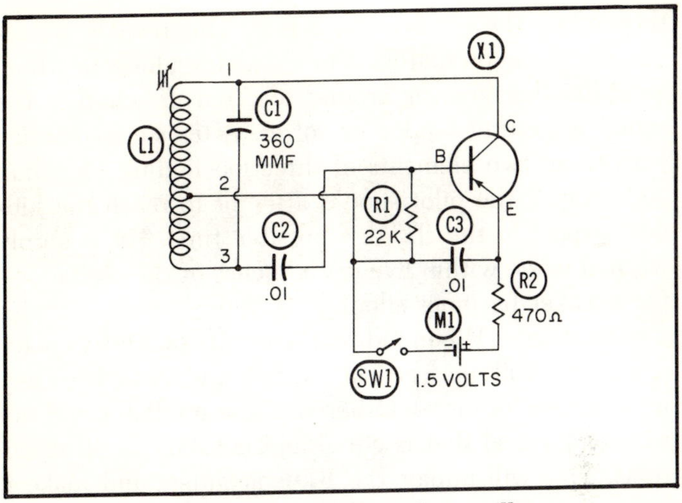

Fig. 13-5. Schematic of the oscillator.

table. Hold the receiver about five feet away and rotate it until the rushing sound stops. You are at a null. Observe the position of the receiver, and note which part of the case points toward the oscillator. This reference point is always used to find the direction of the oscillator.

Walk completely around the oscillator. As you reach a null, observe how the reference point on the case is always aimed toward the oscillator.

Next is the problem of the double null. As the receiver is rotated, you will find two points where the signal disappears. The remedy is to rotate the receiver until you hear a strong signal, and walk toward the oscillator. The hissing sound will become louder and louder, revealing the desired direction. If you walk away from the oscillator, the hissing will become weaker, indicating the wrong direction.

The first step is to hide the oscillator while the "pilot" is out of the room. Upon his return he must then rotate the receiver for a null. This reveals the direction of the oscillator. While walking along this line, he shifts the receiver around to discover whether the signal is getting louder or softer. After some practice with these two elements of direction finding (heading and range), the pilot should attempt to reach the hidden airport in the shortest possible time. He is amply warned when within five or six inches of the oscillator—the speaker becomes silent.

A final note: When hiding the oscillator, always place it with the coil in a vertical position and away from any large masses of metal. Otherwise, the oscillator will not radiate a signal that is equal and constant in all directions. This will impair the RDF accuracy and make it hard to pinpoint the oscillator.

PARTS LIST

Item No.	*Quantity*	*Description*
X1	1	Transistor, pnp, germanium, rf (CK722, TR-05, HEP253, ECG100, SK3004, GE-1, HF20M).
C1	1	360-mmf disc capacitor.
C2, C3	2	.01-mfd disc capacitors.
R1	1	22K, ½-watt resistor.
R2	1	470-ohm ½-watt resistor.
SW1	1	Switch (see text).
L1	1	Transistor loopstick, 3-tap coil (Superex VLT-240 or equivalent).
M1	1	Size A flashlight cell.
	1	2½" square piece perforated board.
	1	Fahnestock clip (for switch).
	1	6-32 machine screw and nut.
	1	Knob (for tuning coil).
		Hookup wire.

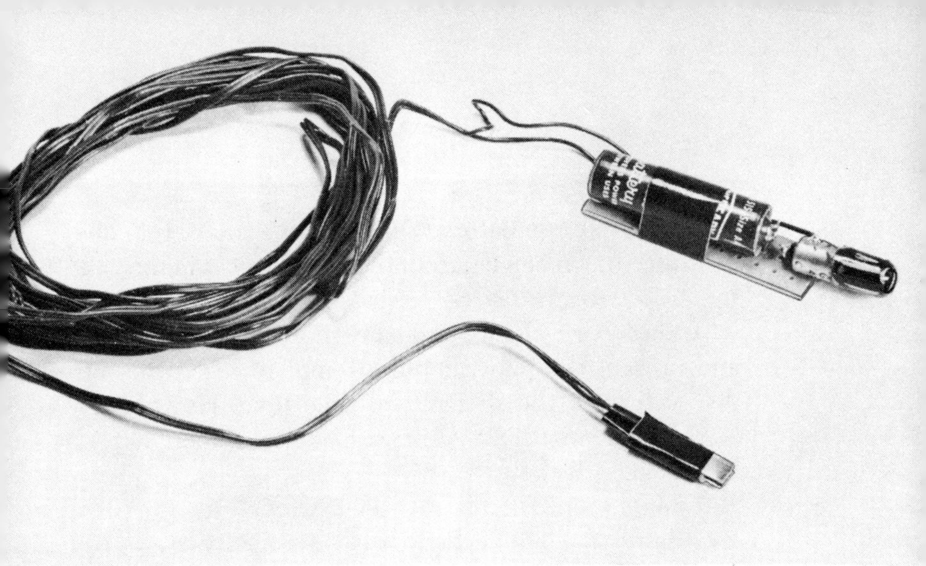

14. MIND-READING ACT

It is almost axiomatic that the best tricks are the simplest. This applies to the following illusion, in which you display remarkable mental feats that lead the audience to believe you are endowed with telepathic powers. The twist, of course, is electronic, in the form of a well-concealed signaling system shown above. Applied with the proper amount of misdirection (the magician's greatest ally), it is highly effective entertainment.

Two accomplices perform the "mind reading act." One is seated with his back to the audience, while the other requests objects from the group. These may be Social Security or currency numbers, or any written information. The performer with his back to the audience then "reads" his partner's mind and calls out the information. If someone complains that the performers are conversing in code, the act can be repeated in complete silence! This will really baffle the heckler.

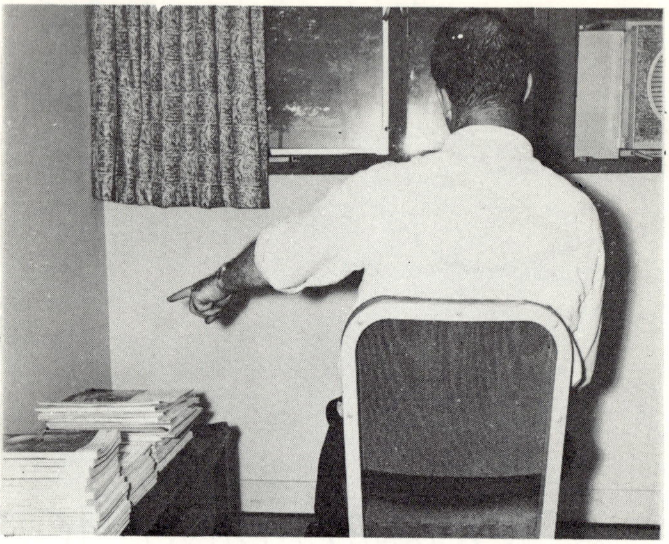

Fig. 14-1. The "Mind Reader" sitting in a chair, and pointing to the signal light.

HOW IT WORKS

The electronic apparatus consists of a specially constructed switch and signal lamp. As you will soon see, the success of the system depends a lot on proper concealment. The switch is thin enough that it can be slid under a rug and leave no revealing bulge or outline. Even stepping on it repeatedly will cause no damage. At the other end of the two-wire cable is a small battery and pilot lamp. Hidden from general view, it is visible only to the performer who is seated. The glow from the bulb is just bright enough for him alone to see, and is further confined to a narrow slit on the bulb covering.

In Fig. 14-1 the performer is pointing toward the signal source. The bulb is placed behind the low table, out of the audience's sight. During the performance it is

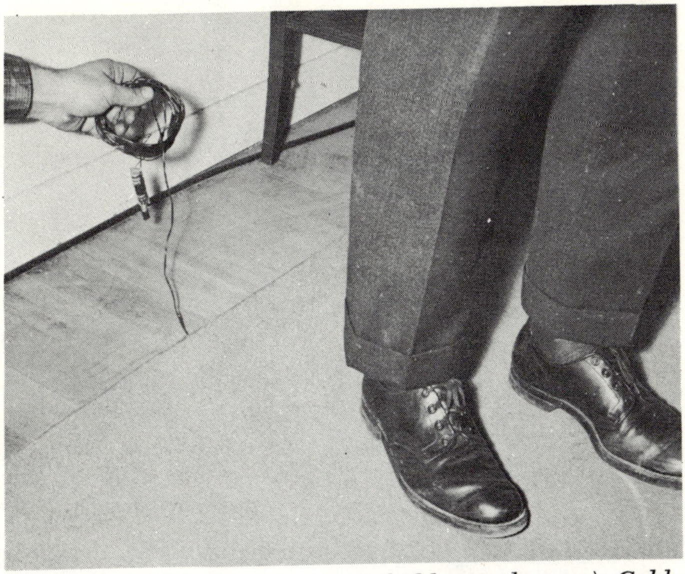

Fig. 14-2. Installation of switch (hidden under rug). Cable is run to a hidden point near other performer.

extremely important that the seated person keep his head facing forward. Shifting his head in the bulb's direction suggests he is looking at something (which he is!). The best technique is to move only the eyes. The audience sees only the back of his head and believes he is looking straight ahead at all times.

The switch operated by the accomplice is shown being installed in Fig. 14-2. It is slid under the rug and the sole of the accomplice's right shoe placed above it. The hand in the photo is stringing the connecting wire over to the chair position. (The wire will be completely hidden under the rug.)

The barest pressure of the performer's shoe is enough to operate the switch. This action is so slight that the audience will not be aware of any leg movement. Since the exact pressure point is important, it is helpful to

Fig. 14-3. First step in constructing switch. Note how wire leads are soldered to copper strips.

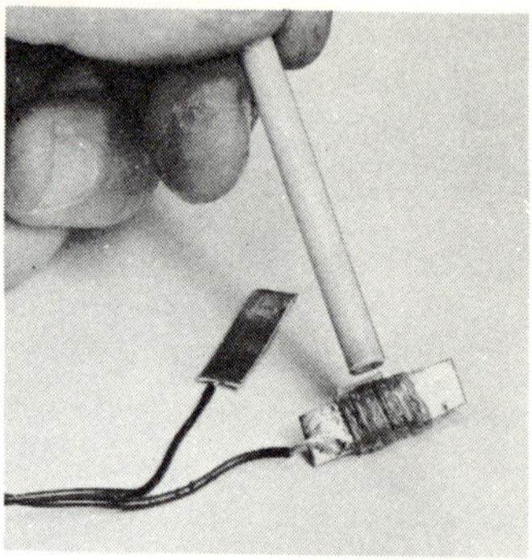

Fig. 14-4. Second step in switch construction. Rubber band is wrapped around one strip to keep copper strips separated.

place a tiny snip of paper on the rug. (This marker is known only to the performer, of course.) Then he may walk freely around the room and be assured that he will always place his foot in the correct position. If the rug has a pattern, use some point in the design instead of the paper.

CONSTRUCTION

Building the switch is a three-step process. First cut two copper strips ¼″ × 1″. Then solder two ends of a long, paired wire to them, as in Fig. 14-3. Next cut a

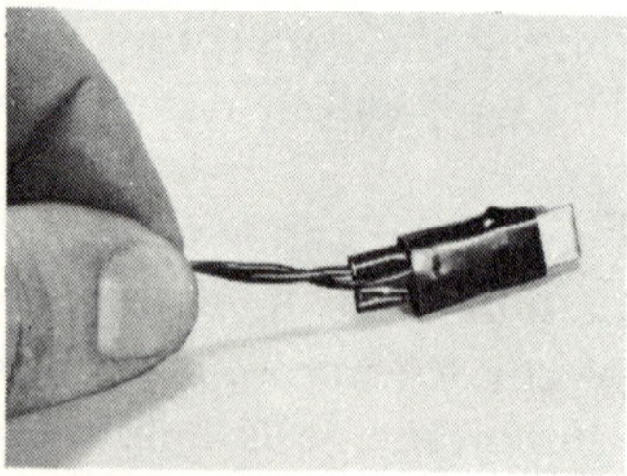

Fig. 14-5. Appearance of the completed switch.

Fig. 14-6. *The completed signal unit.*

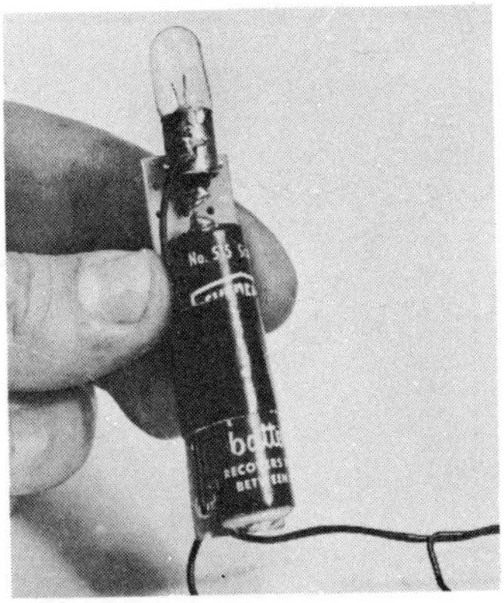

rubber band in two, and wrap a layer about three-fourths of an inch down one strip. This imparts springiness to the switch. Apply glue to hold the rubber in place. The remaining one-fourth inch of bare copper acts as a contact point. Fig. 14-4 shows this step.

Finally, place the other copper strip atop the first and wind two turns of electrical tape around them, as shown

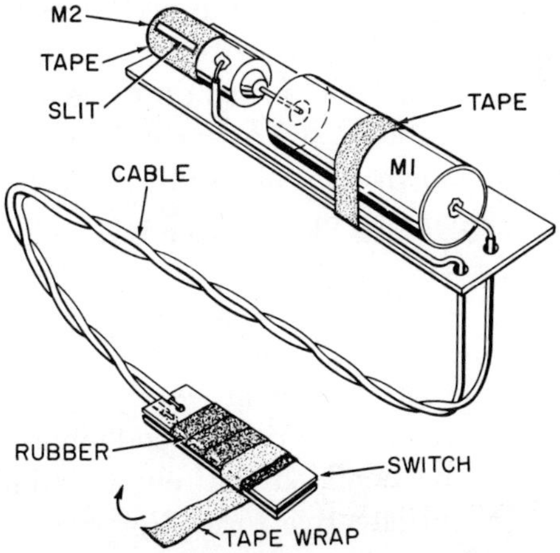

Fig. 14-7. Pictorial diagram of the "Mind Reader."

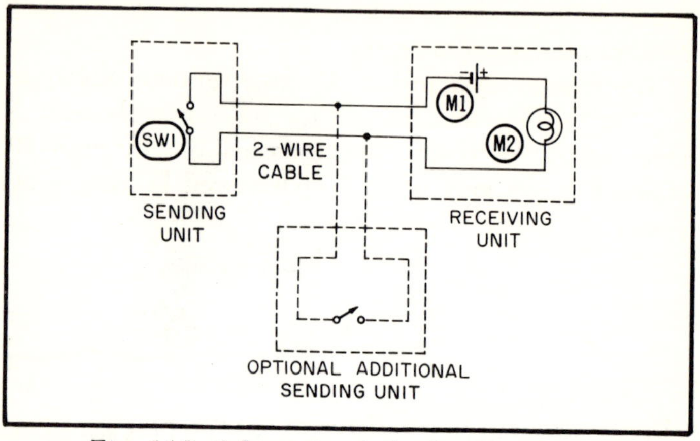

Fig. 14-8. Schematic of the "Mind Reader."

in Fig. 14-5. For smoother switch action, the *unsoldered* surfaces of the copper strips should face the inside.

The other end of the wire connects to the signal assembly consisting of a Size A *Penlite* cell and a lamp supported by a small board. (See Fig. 14-6.) Wrap the lamp with tape except for a narrow slit. This leaves just enough light to be recognizable by the performer. (See Fig. 14-7.) The schematic of the complete circuit is given in Fig. 14-8. To add further mystery to the mind-reading act, additional switches can be constructed and connected as shown. This allows the performer to signal his partner from two or more points.

CODES

Now that the performers are equipped with a means of clandestine communications, a code must be devised. The act may be confined entirely to numbers in its simplest but still effective form. For example, suppose someone in the audience gives the standing performer a series of numbers. These may be sent by pulsing the rug switch once for *1*, twice for *2*, etc. A pause signifies another number coming up.

The number system has the disadvantage of requiring too much transmission time for numerals over *5*. One remedy is to use Morse code for *6* through *10*. Number *6* is a long pulse (–) followed by four short ones (. . . .). The others are:

7 — — . . .
8 — — — . .
9 — — — — .
0 — — — — —

If you are interested in making the mind-reading act a regular part of your repertoire, you and your accomplice should learn the complete Morse code. In this way, both of you will be able to send and receive worded messages. As they say in show biz, you'll have the audience "eating out of your hand"!

PARTS LIST

Item No.	*Quantity*	*Description*
SW1	1	Switch (see text).
M1	1	Size A Flashlight cell.
M2	1	No. 49 pilot lamp.
	2	¼″ × 1″ copper strips.
	1	Rubber band.
	1	½″ × 2″ piece hardboard.
		2-wire cable.
		Electrical tape.

15. TREASURE HUNT

In attempting to find the hidden treasure in the shortest time, the player soon discovers his biggest obstacle is—people. The "treasure" he tries to track down will always be concealed inside the clothing of someone in the group. Pinpointing its location is a sure-fire way of giving a hilarious lift to a party.

The operation of the game is shown above. The player is manipulating an electronic gadget near his prime suspect; the fellow with his hands in the air. He sweeps his victim's body for a tell-tale signal. If none is forth-

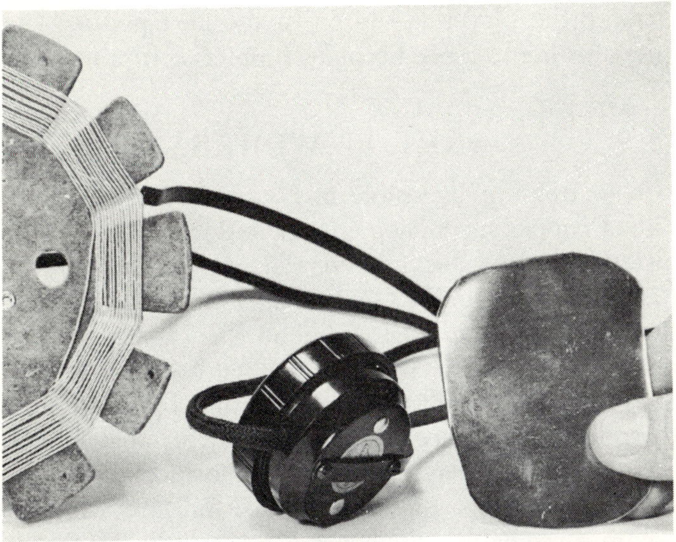

Fig. 15-1. The copper disc held by the hand is the "Treasure." The signal is heard in the headphone.

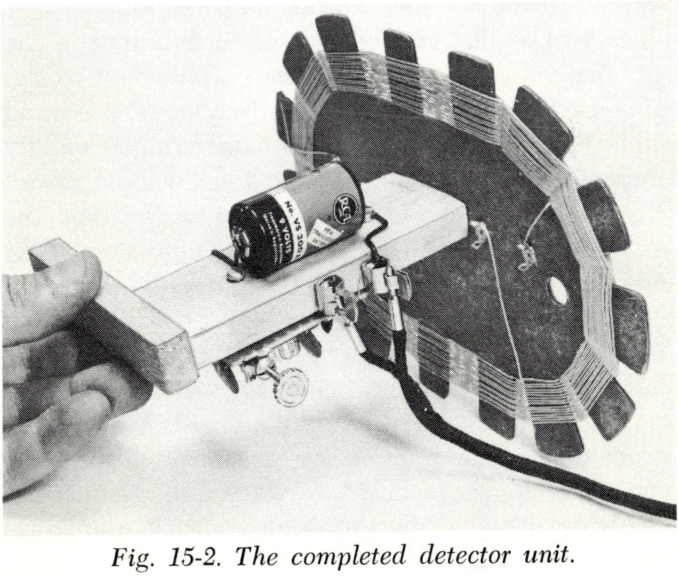

Fig. 15-2. The completed detector unit.

coming, the player moves on to the next person. The quest for the treasure becomes funnier as it progresses.

HOW IT WORKS

The "treasure" is visible in Fig. 15-1. It is simply a disc of copper small enough to be hidden under a shirt, inside a shoe or stocking, etc. The device used to find the disc appears in Fig. 15-2. It is an electronic detector that sounds a tone in an earphone when in the vicinity of the copper disc. The pickup range is only a few inches, but this is sufficient to sense the copper through the nonmetallic objects (such as clothing) that hide it.

The circuitry may be understood by looking at the schematic in Fig. 15-3. Two transistors are wired as separate oscillators. Each generates a radio frequency of approximately 500 to 600 kc. These frequencies are matched perfectly by adjusting tunable coil L2 and listening to the earphone during construction.

The two oscillators feed part of their output to the earphone via the transistor emitters. Here the two signals mix. If they are identical in frequency, no sound will be heard. However, if they are operating at slightly different frequencies, an audible tone will be heard. This tone is the result of heterodyning, or "beating" the two frequencies together. Say one oscillator is on 600 kc and the other is on 601. The difference signal is 1 kc, or 1,000 cycles. Such a tone is audible to the human ear.

At the beginning of the game, the oscillators are matched by zero-beating them. This is done by slowly rotating the coil knob of L2 until a tone is heard in the earphone. Further rotation lowers the pitch until the tone can no longer be heard. Now the radio frequencies are identical; this is the zero-beat point. Continued rotation in the same direction causes the tone to be heard

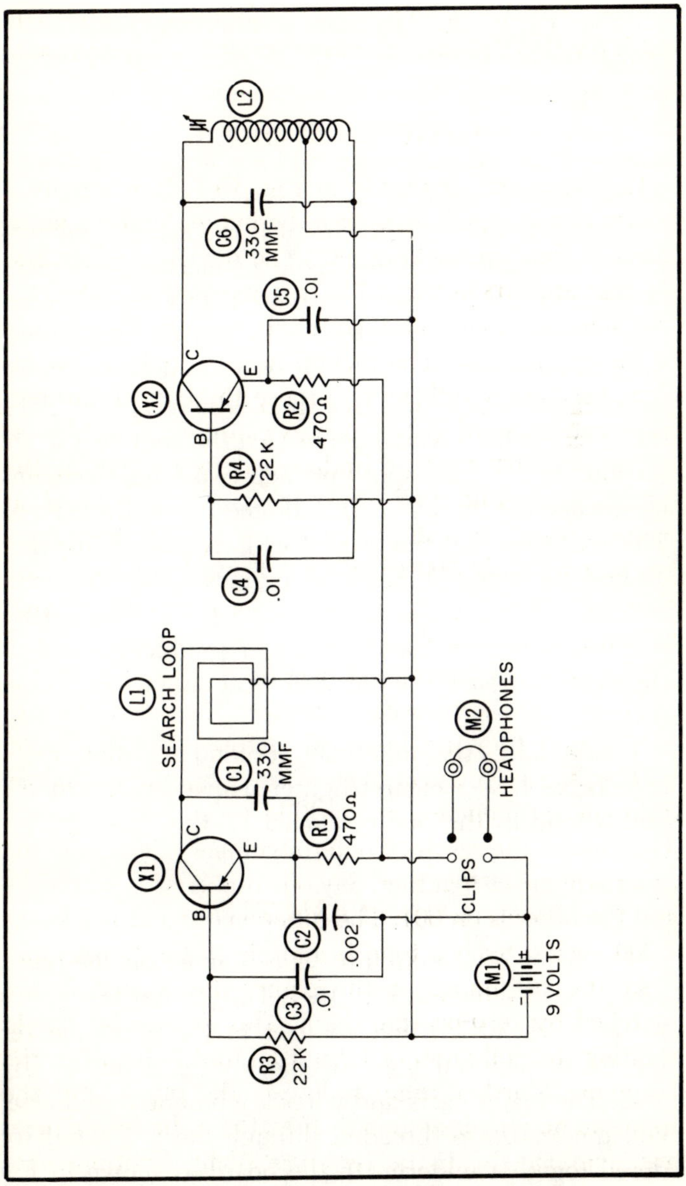

Fig. 15-3. Schematic of the "Treasure Hunt."

again as a low pitch that rises until it can no longer be heard. We are interested in the zero-beat setting, where no tone is heard. Then, as the search coil (L1) nears the copper disc, its inductance changes. This throws one oscillator off-frequency, and a warning tone is generated in the earphone.

Another feature of the circuit is the use of "pulling," which ordinarily is undesirable in electronic devices. "Pulling" occurs whenever two oscillators are in a common circuit. One oscillator "pulls" on the other until their frequencies are the same. We use it here to keep the unit stable and thus free from false responses. After the device is built, you'll notice that the zero beat in the earphone (the dead spot between two tones) remains, even when the tuning knob is rotated two or three turns. This indicates that the oscillators have pulled together. To find the strongest lock-in point, move your open hand toward and away from search coil L1. The best setting occurs where your hand no longer produces a tone in the earphone. Now the unit will react only to the copper disc.

Incidentally, the unit may respond to other metal objects, such as a cigaret lighter. However, the tone is not as loud as the one produced by the "treasure."

CONSTRUCTION

The heart of the circuitry is contained on the perforated board shown in Fig. 15-4. The pointer is pointing to the leads which will connect to the earphone terminals. A pictorial view of the unit is given in Fig. 15-5. The major parts and wires are mounted as shown, and the leads are threaded through the holes and soldered together underneath the board as shown in Fig. 15-6. Note the position of the tuning knob at the upper

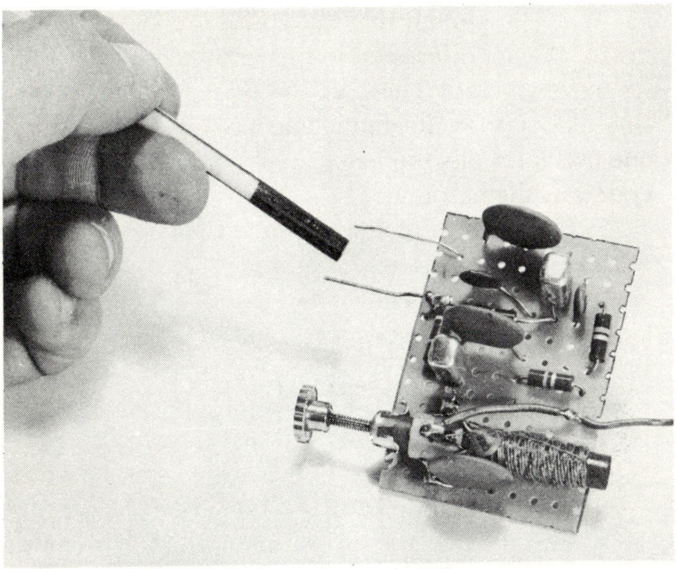

Fig. 15-4. Top view of the completed circuit board.

right; the two wires next to it go to the earphone terminals. The pointer shows the negative battery lead.

After completing the board, fasten it to the two-piece wooden handle and connect the clips for the earphone. The two clips for the earphone act as jacks for its pin tips (if the 'phone has tips). Clips must be used, even with units that terminate in wire leads, because the earphone serves as the on-off switch. No power is supplied until the 'phone is clipped into position.

Fasten the batteries to the handle by its leads. Simply drive a wood screw over the leads and into the wood handle.

Next mount the large search coil, which is actually a standard replacement loop for most AM table radios. It is important to select the correct terminals on the loop for the two leads running to the circuit board. The loop in the model shown had four solder lugs.

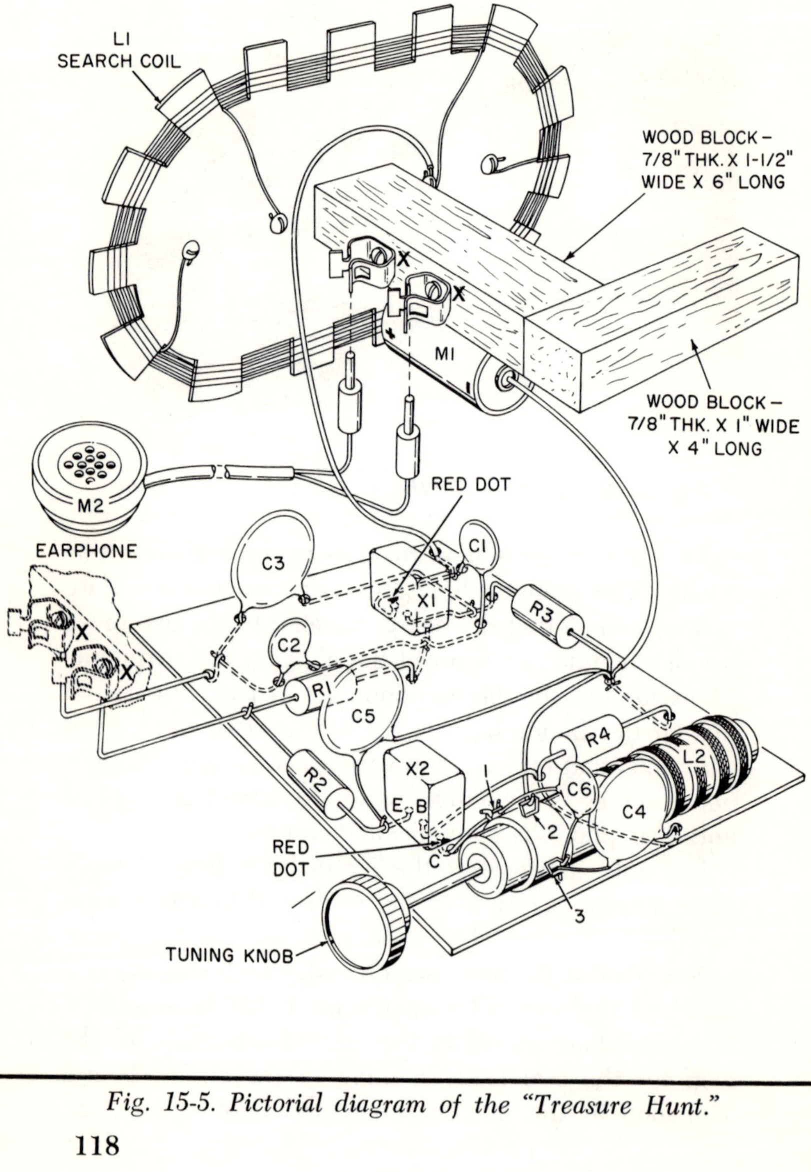

Fig. 15-5. Pictorial diagram of the "Treasure Hunt."

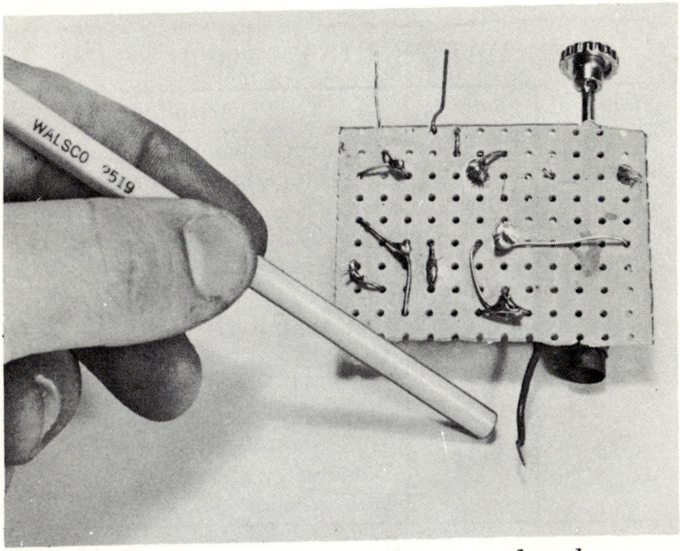

Fig. 15-6. Underside of the circuit board.

(Some have only three.) Finding the right pair is done with an ohmmeter; use the two loop terminals displaying the most resistance (about 10 ohms).

PARTS LIST

Item No.	*Quantity*	*Description*
R1, R2	2	470-ohm, ½-watt resistors.
R3, R4	2	22K, ½-watt resistors.
C1, C6	2	330-mmf disc capacitors.
C2	1	.002-mfd disc capacitor.
C3, C4, C5	3	.01-mfd disc capacitors.
L1	1	Antenna loop (broadcast type).
L2	1	Tapped transistor antenna coil (Superex VLT-240 or equivalent).
M1	1	9-volt battery.

(continued)

PARTS LIST (cont'd)

Item No.	*Quantity*	*Description*
M2	1	1000-ohm dynamic earphone.
X1, X2	2	Transistors, pnp, germanium, rf (CK722, TR-05, HEP253, ECG100, SK3004, GE-1, HF20M).
	1	2½″ × 1¾″ perforated board.
	1	Copper disc.
	1	⅞″ × 1″ × 4″ wood strip.
	1	⅞″ × 1½″ × 6″ wood strip.
	2	Clips (for earphone).
	1	Tuning knob.
		Wood screws.
		Hookup wire.

16. THE "MUSITRON"

Instruments that produce electronic music were once considered laboratory curiosities. Lately they have gained considerable recognition. The unit described here is based on audio oscillation, which is one of the key principles used in the creation of sound by purely electronic means. Dubbed the "Musitron," the device is capable of a wide range of sounds and tones. It will be of special interest to persons with musical inclinations, as well as simply a diverting form of amusement.

The above photo shows how the "Musitron" is played. The left hand presses a button that starts the tone, while the right operates a handle to determine its pitch. When the handle is in its maximum clockwise position, the

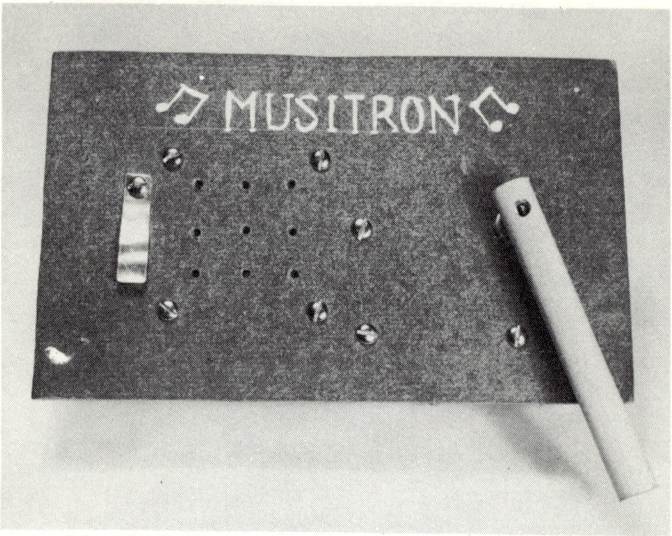

Fig. 16-1. Front view of the completed board.

sound is very low in frequency, or pitch—about 30 or 40 cycles per second. The pitch rises as the handle is turned in the other direction, and reaches a high of several thousand cycles at the other end of rotation. This gives the instrument a range of several octaves, enough to play almost any tune.

With practice you can impart a variety of melodious effects. Simply pressing the button each time the handle is moved will produce an uninteresting, though accurately pitched, rendition. Vibratro and tremolo can significantly heighten the tone quality. Both techniques create a rapid wavering of sound which tends to remove the "flatness" of the notes. The result makes the device more a musical rather than an electronic instrument.

First consider vibrato. This is produced by rapidly shifting the *frequency* of the tone. (A violinist does this by moving his fingers back and forth over the strings.) In the "Musitron," vibrato is achieved by a slight quiv-

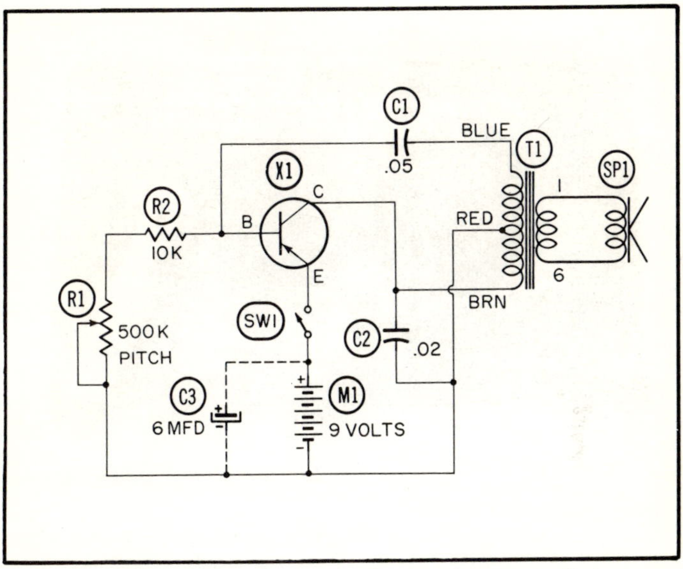

Fig. 16-2. Schematic of the "Musitron."

ering of the hand holding the pitch handle shown at the right in Fig. 16-1.

Tremolo varies the amplitude or strength, of the tone. Note the holes on top of the board; they allow sound from the speaker to come through. When the hand is fluttered just above them, a tremolo effect occurs as the hand alternately blocks and unblocks the sound. The tone button on the left (Fig. 16-1) is pressed with the little finger of the left hand, freeing the other four for the tremolo. As with any musical instrument, a certain amount of practice will be needed.

HOW IT WORKS

The schematic of the unit appears in Fig. 16-2. The unit is a one-transistor audio oscillator. The pitch is altered by the setting of potentiometer R1. Transformer

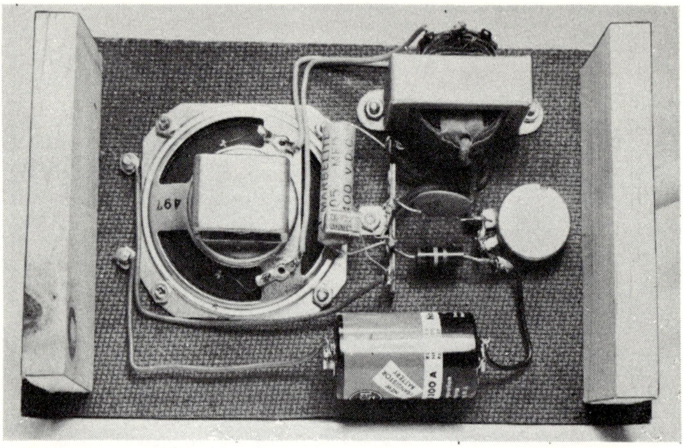

Fig. 16-3. Underside of the completed board.

T1 acts as part of the oscillator's tuned circuit and also matches the audio energy from the transistor to the speaker.

C3, shown in dotted lines, is optional and is not shown in the photos or drawings in this chapter. As the battery ages, its internal resistance goes up. Eventually a harsh humming sound will be heard on the high tones when the handle is fully counterclockwise. When this happens, install C3 (an electrolytic capacitor) across the positive and negative terminals of the battery. It will prolong the usefulness of the battery by bypassing the effects of internal resistance. Be sure to attach the positive lead of C3 to the positive terminal of the battery, and the negative lead to the negative terminal.

CONSTRUCTION

Cut the board to 8″ × 4½″. Mount all major components (except the speaker) as shown in Fig. 16-3. Drill nine holes in the board, to let the sound through. Now

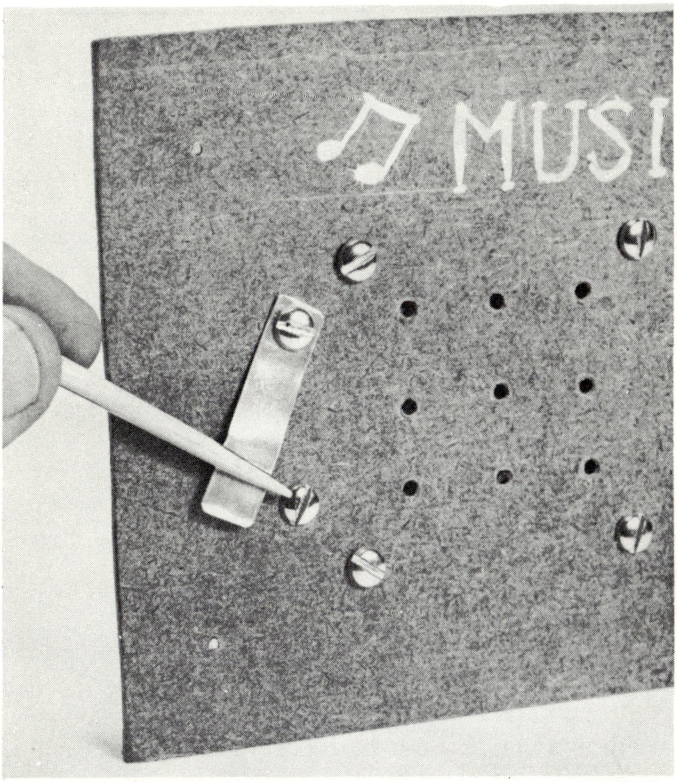

Fig. 16-4. Closeup of the left side of the board, showing the switch construction and the holes which allow sound from speaker to pass through board.

you can mount the speaker, securing it with machine screws and nuts.

Make the button from a strip of copper about 1½″ × ⅜″. Secure one end of the strip with a machine screw and nut (toward the top edge of the board). Place another machine screw under the lower end of the copper strip as shown in Fig. 16-4 to serve as the other switch contact. When the button is depressed, the free end of the copper strip should strike the screw head

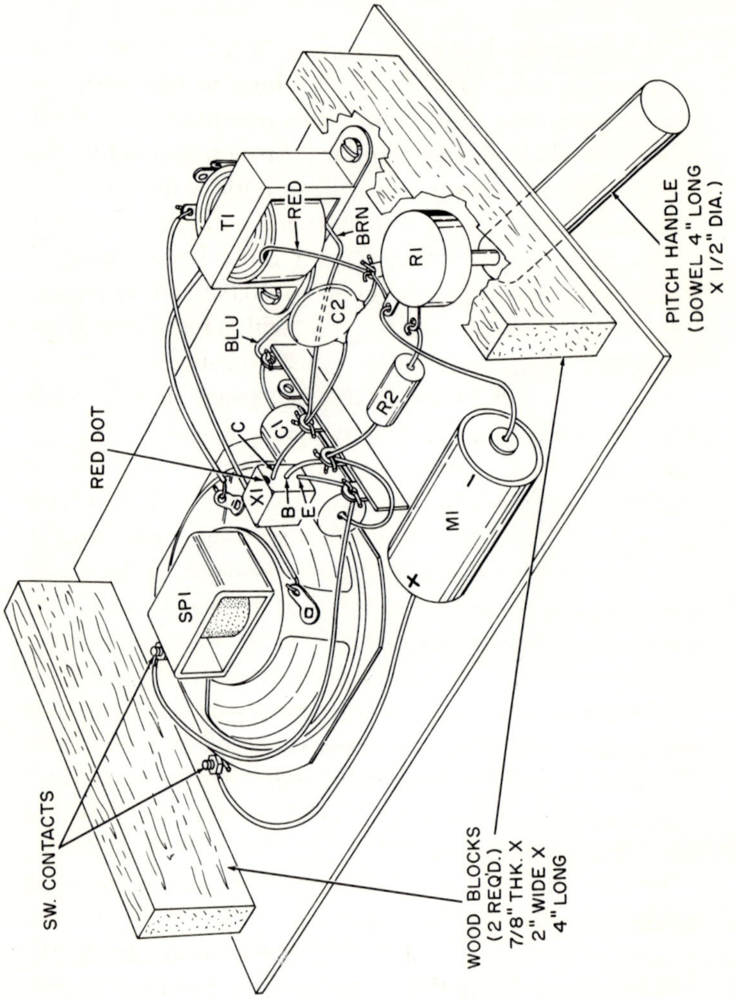

Fig. 16-5. Pictorial diagram of the "Musitron."

below. For connections to the switch contacts, solder wires to the threads of the screws from underneath the board.

Make the pitch handle from a ½-inch wood dowel four inches long. Drill a 7/32-inch hole in one end, and gently press the dowel onto the potentiometer shaft. Since the shaft is the standard ¼-inch diameter, the slightly smaller dowel hole will provide a firm fit.

Many speaker and transformer combinations will work. However, the transformer must be a push-pull type with approximately 4,000 to 10,000 ohms impedance in its primary winding. It will have three wires emerging from one side (customarily color-coded brown, red, and blue). Wire as shown in Fig. 16-5. The secondary winding and speaker must have the same impedance (often 3.2 ohms, voice coil). If the transformers given in the Parts List are used, you have some flexibility in the choice of a speaker. For example, you can use one of the popular transistor replacement speakers, which are often rated at 10 or 11 ohms. The transformer has several lugs, or taps, which provide a variety of voice-coil impedances. In the model shown, a 10-ohm speaker was installed and lugs 1 and 6 gave the best match. With other speakers, you can determine the best secondary connection very easily. Just touch the speaker leads to each lug until you find the pair that gives the loudest sound.

Mount the remaining components and connecting leads as shown in Figs. 16-3 and 16-5. Make sure the positive (+) terminal of the battery is connected to the switch contact and the negative (−) terminal is connected to potentiometer R1.

Finally, attach the two wood strips to support the unit so it will not rest on the components when being played.

PARTS LIST

Item No.	*Quantity*	*Description*
X1	1	Transistor, pnp, germanium, rf (CK722, TR-05, HEP253, ECG100, SK3004, GE-1, HF20M).
R1	1	500K, audio-taper, carbon-type potentiometer.
R2	1	10K, ½-watt resistor.
C1	1	.05 disc or paper capacitor.
C2	1	.02 disc or paper capacitor.
C3	1	6-mfd, 15-volt electrolytic capacitor (optional).
T1	1	Audio output transformer (Merit A-2900, Stancor A-3856, Thordarson 24S04, Triad S-51X, or equiv.).
SP1	1	2- or 3-inch speaker.
SW1	1	Switch (see text).
M1	1	9-volt battery.
	1	8″ × 4½″ piece hardboard.
	1	4″ length of ½″ wood dowel.
	1	5-lug terminal strip.
	1	⅜″ × 1½″ copper strip.
	2	⅞″ × 2″ × 4″ wood strips.
	9	6-32 machine screws.